HANON

娃衣缝纫书

（日）藤井里美　著

王春梅　译

辽宁科学技术出版社

沈阳

目录

细节调整

作者：藤井里美
摄影：葛贵纪、田中麻子、玉井久义
数码化：久助 YUKARI
编辑：铃木洋子
设计：田中麻子
其他：

be my baby! Cherry：M size
株式会社 MIYUKING
www.doll-house-web.com

JERRY BERRY：S size
JERRYBERRYs.com
jerryberrys.net

POPPING：S size
KUKUCLARA
kukuclaradoll.com

ruruko：S size
momoko(p.18,21)：L size
Pet works doll 事业部
www.petworks.co.jp/doll/

U-noa Quluts Light：L size
株式会社 SEKIGUCHI
www.sekiguchi.co.jp

这本书是我的第二本书，同样介绍了如何缝制洋娃娃尺寸的小洋装。与前一本相比，本书更加简单，以更适合新手进行缝制的方法为基础，详细讲解了洋装主体、领子、袖子以及褶皱等部位的详细处理方式。

各位读者可以尽情地把基本款与调整款组合在一起，享受搭配的乐趣。

本书中按照 1/6 的尺寸，附带了 3 种类型的型纸。

S 码，适用于 ruruko、JERRY BERRY、Popping 等

M 码，适用于 Neo Blythe 和 b.m.b.Cherry 等

L 码，适用于 U-noa Quluts Light 和 momoko 等

虽然本书中均使用缝纫机进行缝制，但也完全可以改为手工缝制。缝制方法多种多样，请选择最适合自己操作的方法吧。希望能给您提供一定的参考。

我希望人人都能体会到搭配的乐趣，所以特意准备了多款长短不一的裙摆和围裙款式。另外，还有大小正合适、可套在外面的外套、夹克衫等服饰。当然，你也可以把这些单品和第一本娃衣缝纫书中介绍的洋装、袜子、靴子和毛绒玩偶等搭配在一起。

请把 HANON 风格带入您家的娃娃衣柜吧。

如果这本书能使你对娃娃们倾注更多的关爱，我将不胜欣喜。

HANON

藤井里美

This book is the second edition of DOLL SEWING BOOK HANON,
which is for making doll-sized clothes. Based on a simpler and easier pattern
than the first one, you can enjoy your bodice, collar, sleeve, skirt, etc.
by combining your favorite arrangements.

S size is for ruruko, JERRY BERRY, Popping
M size is for Neo Blythe, b.m.b.Cherry
L size is for U-noa Quluts Light, momoko.

The book describes using sewing machine,
but I think you can also sew by hand-sewing reverse stitching.
There are various sewing methods, but please try to make clothes
by a sewing method that suits yourself. I hope this book helps you.

In order to let you know the enjoying of coordination,
I have prepared several length of skirts and aprons pattern.
Please enjoy coordination with clothes, socks, shoes, and stuffed toys
that are published in the first edition.

Please add the HANON style to your doll's wardrobe,
I would be glad if I could help you more shower your dolls with love.

HANON
Satomi Fujii

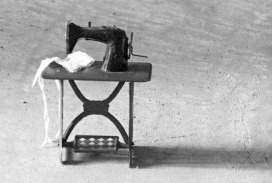

M 码基本款收腰连衣裙（细节调整）、连衣围裙（细节调整）和帽子

M 码基本款收腰连衣裙（细节调整）和手袋

M 码基本款收腰连衣裙（细节调整）和活领

M 码基本款 A 字连衣裙（细节调整）

M 码夹克衫、基本款衬衫（细节调整）灯笼裤、帽子和手袋

M 码基本款 A 字连衣裙（细节调整）

M 码基本款衬衫（细节调整）、活领和帽子 / M 码外套和帽子

M 码基本款 A 字连衣裙（细节调整）、连衣围裙和手袋

M 码基本款衬衫（细节调整）、灯笼裤／M 码基本款衬衫（细节调整）、荷叶裙和连衣围裙（细节调整）

M 码基本款 A 字连衣裙（细节调整）和帽子

S 码基本款 A 字连衣裙（细节调整）

（上）S 码基本款收腰连衣裙（细节调整）/（下）S 码外套和基本款收腰连衣裙（细节调整）

S 码夹克衫和荷叶裙

S 码基本款衬衫（细节调整）和荷叶裙

S 码基本款 A 字连衣裙（细节调整）和连衣围裙／L 码基本款收腰连衣裙（细节调整）和连衣围裙

S 码基本款衬衫（细节调整）、灯笼裤、帽子和手袋

S 码基本款收腰连衣裙（细节调整）

S 码基本款收腰连衣裙（细节调整）和帽子
L 码基本款衬衫（细节调整）、荷叶裙和帽子

L 码基本款 A 字连衣裙（细节调整）和活领（细节调整）/ L 码基本款 A 字连衣裙（细节调整）

L 码基本款衬衫（细节调整）连衣围裙

L 码夹克衫、灯笼裤和帽子／L 码外套、基本款衬衫（细节调整）灯笼裤和帽子

在开始缝制娃娃的洋服之前，先准备好工具吧。

有些平时缝纫时几乎不会用到的工具，在缝纫小尺码的娃娃洋装时却十分有用。

绣花丝带 *Embroidery Silk Ribbon*
用 3.5mm 宽的丝带来刺绣蝴蝶结，质地柔软，便于使用，而且颜色丰富。

绣线 *Cotton Embroidery Floss*
使用 DMC 的 25 号绣线。

拆线刀 *Seam Ripper*
缝纫线不端正的时候，可以用拆线刀把绣线清除干净，然后重新缝纫。

镊子 *Tailor's awl*
手工用的小镊子，能顺利地把小巧的布料翻面。

绣线剪 *Thread Scissors*
无论动手缝制还是使用缝纫机，都少不了用绣线剪来剪断绣线。

顶针 *Thimble*
在刺绣或扦边时使用。

锥子 *Tailor's Bodkin*
用于布料翻到反面时顶出边角，用缝纫机时也可以用于压住布料。

缝纫剪 *Dressmaking Scissors*
剪口锋利，可以完成精细操作的小号剪刀。

缝纫线 *Sewing Thread*
手工也好，缝纫机也罢，本人一直钟爱"TicTic PREMIER"的缝纫线。

布胶 *Fabric Glue*
临时固定的时候，选择布胶。要长期固定的时候，建议使用皮革用手工胶。

布料锁边液 *Fray Stopper*
布料裁剪开以后，涂在切口处防止飞边。

裁缝粉笔 *Tailor's Chalk*
对于质地纤薄的布料，使用不易渗色的铅笔；对于质地厚重的布料，使用极细的粉笔；而对于深色布料，则使用白色粉笔。

蕾丝类 *Laces*
书上使用的是复古蕾丝。如果您觉得新蕾丝的颜色过于靓丽，与服饰不够匹配，可以选用草木染或红茶染的蕾丝，以自己喜欢的颜色为宜。

别针 *Snaps*
本人使用有 5mm 小圆头的别针。

缝纫针、裁缝别针、丝绸别针、直尺
Handsewing Needles, Dressmaker Pins, Silk Pin, Ruler

Basic A-line Dress
基本款 A 字连衣裙

只需 3 张型纸就能制作，即使是缝纫新手也能轻松上手的简洁款连衣裙。领子、袖子、褶皱、主体
完成以后，还可以根据自身喜好进行细节部分的调整。

锦纶 （大身）	S	20cm×32cm	锦纶 （袖子）	S	12cm×22cm	内衬 （尼龙）	S，M，L 7cm×7cm
	M	20cm×38cm		M	12cm×24cm		
	L	25cm×40cm		L	15cm×28cm		
锦纶 （领子）	S	4cm×20cm	锦纶 （下摆布料）	S	4cm×60cm	按扣	S，M，L×2 组
	M	4cm×20cm		M	4cm×60cm		
	L	4cm×23cm		L	4cm×68cm		

基本款 A 字连衣裙

1

配合型纸，裁剪好各部分的布料，用布料锁边液对布料边缘进行处理。
→大身的细节调整，请参考 p.79~83

2

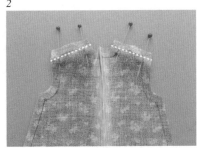

把大身前片和大身后片的反面对齐，缝合肩部。

3

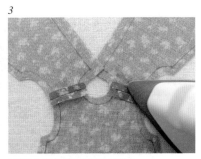

用熨斗把缝头处的布料分开熨平。
→如果有领子，请参考 p.84~88

4

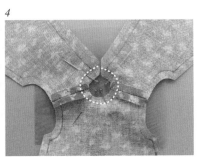

准备一枚 7cm×7cm 大小的毛裁内衬料，与大身片反面对齐，然后缝合一圈领子。

5

如图，裁剪内衬的材料，然后沿着领子缝头剪开细小的切口。小心，不要剪断缝头处的针脚。

6

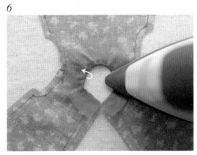

翻回反面，用熨斗熨烫平整。

7

从反面把缝头处的布料斜着折进来，一直折到略低于开口止位的地方，然后用熨斗熨烫平整。

8

沿着略低于开口止位、领口、略低于开口止位处缝合。
→如果有袖子，请参考 p.89~95

9

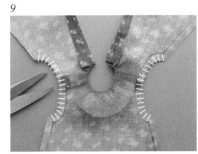

在袖筒的缝头处，剪开细小的切口。

---

*Let's enjoy that collars, sleeves and hem arrangements
according to your level.*

1. Arrange the paper templates on the fabric and cut all the sections, then apply fray stopper liquid to all the edges.
[refer to p.79–83 for the front arrangements] *2.* Match the right sides of the front and back by the shoulders and sew.
3. Iron open the seam allowances. [refer to p.84–88 for the collar arrangements] *4.* Cut the dough about 7cm square for lining.
Match the bodice and the lining and sew the neckline. *5.* Cut the lining as shown and cut slit in the seam allowance of the neckline.
Be careful not to cut the stitches. *6.* Turn the lining right side out and iron.
7. Fold the back opening inwards slightly below the opening stop marker and iron flat. *8.* Sew the edges from the stop marker along the
neckline, then to the stop marker. [refer to p.89–95 for the sleeve arrangements] *9.* Cut fine slits into the seam allowance of the armholes.

10

用熨斗把缝头处折过来，然后用布胶临时固定。

11

缝合袖筒。

12

将大身前片和大身后片的反面对齐，缝合两侧。

13

用熨斗把缝头处的布料分开熨平。

14

把下摆的缝头处向内折过来，用熨斗压住。
→下摆的细节调整，请参考 p.96~99

15

缝合下摆。

16

从反面缝合后身的开口，需要从开口止位一直缝合到下摆。

17

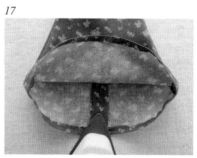

翻回正面，用熨斗把缝头处的布料分开熨平。

18

固定按扣，完成。

10. Fold the seam allowances and apply fabric glue to the seam allowance.   *11*. Sew the armholes.
12. Match the right sides of the front and back bodice sections facing, and sew them together.
13. Iron open the seam allowance.   *14*. Fold the seam allowance of the skirt hem inwards with an iron. [refer to p.96−99 for the hem arrangements]
15. Sew the hem.   *16*. With right sides facing, sew from the stop marker to the hem.   *17*. Iron open the seam allowance.   *18*. Add snaps to the back opening.

Basic Blouse
基本款衬衫

制作方法与 A 字连衣裙基本相同。下摆有无褶边，决定衬衫的整体风格。如果去掉褶边，用荷叶边
来取而代之，就会成为一件完全不同的荷叶边衬衫。请尝试一下不同的搭配组合吧。

锦纶 （大身）	S M L	10cm×24cm 10cm×26cm 11cm×28cm	锦纶 （袖子）	S M L	12cm×22cm 12cm×24cm 15cm×28cm	内衬 （尼龙） 按扣	S，M，L 15cm×15cm S，M，L 2 组
锦纶 （领子）	S M L	4cm×20cm 4cm×20cm 4cm×23cm	锦纶 （下摆布料）	S M L	4cm×36cm 4cm×38cm 4cm×40cm		

基本款衬衫

1

配合型纸，裁剪好各部分的布料，用锁边液对布料边缘进行处理。
→大身的细节调整，请参考 p.79~83

2

把大身前片和大身后片的反面对齐，缝合肩部。

3
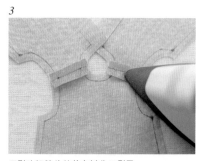

用熨斗把缝头处的布料分开熨平。
→如果有领子，请参考 p.84~88

4

准备一枚 15cm×15cm 大小的毛裁内衬料，与大身片反面对齐，用缝纫别针固定。

5

依次沿着大身后片的下摆、领口、大身前片的下摆处缝合。

6

如图，裁剪内衬的材料，然后沿着领子缝头剪开细小的切口，剪掉边角料。小心，不要剪断缝头处的针脚。

7
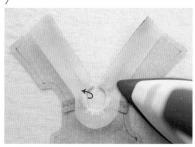

翻回反面，用熨斗熨烫平整。
→如果有袖子，请参考 p.89~95

8
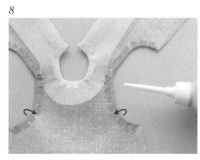

在袖筒的缝头处剪开细小的切口。用熨斗把缝头处折过来，然后用布胶临时固定。

9

压住大身后片的下摆、领口、大身后片的下摆和袖筒，用缝纫机缝合。

It is very easy to make it as Basic A-line dress.
Please enjoy various arrangements.

1. Arrange the paper templates on the fabric and cut all the sections, then apply fray stopper liquid to all the edges.
[refer to p.79−83 for the front arrangements]　*2.* Match the right sides of the front and back by the shoulders and sew.
3. Iron open the seam allowances. [refer to p84−88 for the collar arrangements]　*4.* Cut the dough about 15cm square for lining.
Match the bodice and the lining.　*5.* Sew as pictured.　*6.* Cut the lining as shown and cut slit in the seam allowance of the neckline.
Be careful not to cut the stitches.　*7.* Turn the lining right side out and iron. [refer to p.89−95 for the sleeve arrangements]
8. Cut fine slits into the seam allowance of the armholes. Fold the seam allowances and apply fabric glue.　*9.* Sew the edges as pictured.

10

将大身前片和大身后片的反面对齐，缝合两侧。

11

用熨斗把两侧缝头处的布料分开。
→下摆的细节调整，请参考 p.96~99

12

把下摆的缝头处向内折过来，用熨斗压住。

13

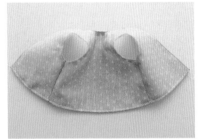

在下摆处，用单股线 2.5mm 宽的针脚缝制，用来
稍后做褶边。
→褶边请参考 p.100

14

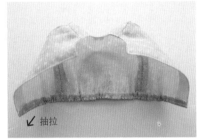

↙ 抽拉

抽拉褶边线，让褶边堆叠起来 (S：11.5cm M：
12cm L：13.5cm)。

15

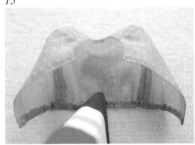

把褶边整理均匀，然后用熨斗熨平。

16

从正面按压，缝合。

17

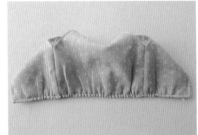

抽出褶边线。

18

在后边开口处固定按扣，完成。

10. Match the right sides of the front and back bodice sections facing, and sew them together.   *11.* Iron open the seam allowance. [refer to p.96−99 for the hem arrangements]   *12.* Fold the seam allowance of the hem inwards with an iron. If you don't want to gather the hem, sew the edges.   *13.* If you want to gather the hem, Use a machine to sew gathering stitches in the seam allowance of the hem. [refer to p.100 for gathering] Make the stitch length 2.5mm and sew one lines on the seam allowance.   *14.* Gather the hem to [S:11.5cm M:12cm L:13.5cm] with the finished line.   *15.* Shape the gathering and iron flat.   *16.* Sew the hem.   *17.* Remove the gather thread.   *18.* Add snaps to complete the blouse.

Basic Darts Dress

基本款收腰连衣裙

通过调整细节，只要对腰部做略微修改，就变成收腰连衣裙。裙摆采用长方形布料，能简单地体验
到变更下摆款式的乐趣。

锦纶	S	15cm×50cm	锦纶	S	12cm×22cm
（连衣裙）	M	15cm×52cm	（袖子）	M	12cm×24cm
	L	17cm×60cm		L	15cm×28cm
锦纶	S	4cm×20cm	内衬	S,M,L	15cm×15cm
（领子）	M	4cm×20cm	（尼龙）		
	L	4cm×23cm			
			按扣	S,M,L	2 组

基本款收腰连衣裙

1

配合型纸，裁剪好各部分的布料，用锁边液对布料边缘进行处理。
→大身的细节调整，请参考 p.79~83

2

把大身前片的腰省翻折过来，缝合。

3

用熨斗把缝头向内侧折好熨平。

4
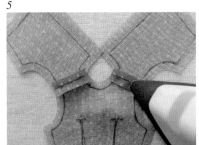

将大身前片和大身后片的反面对齐，缝合肩部。

5

用熨斗把缝头处的布料分开熨平。
→如果有领子，请参考 p.84~88

6
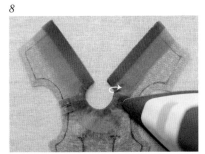

准备一片 15cm×15cm 大小的毛裁内衬料，与大身片反面对齐，沿着后面开口的下摆、领口、后面开口的下摆处缝合。

7

如图，裁剪内衬的材料，然后沿着领子缝头剪开细小的切口。小心，不要剪断缝头处的针脚。

8

翻回反面，用熨斗熨烫平整。

9

压住后面开口的下摆、领口、后面开口的下摆，用缝纫机缝合。
→如果有袖子，请参考 p.89~95

The skirt has a rectangular pattern,
so you can easily enjoy the arrangement.

1. Arrange the paper templates on the fabric and cut all the sections, then apply fray stopper liquid to all the edges. [refer to p.79−83 for the front arrangements]   2. Fold and sew the darts.   3. Fold the seam allowances of each darts inward and iron.
4. Match the right sides of the front and back by the shoulders and sew.   5. Iron open the seam allowances. [refer to p.84−88 for the collar arrangements]   6. Cut the dough about 15cm square for lining. Match the bodice and the lining and sew as pictured.
7. Cut the lining as shown and cut slit in the seam allowance of the neckline. Be careful not to cut the stitches.
8. Turn the lining right side out and iron.   9. Sew the edges as pictured. [refer to p.89−95 for the sleeve arrangements]

10

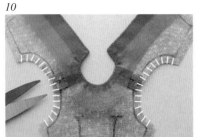

在袖筒的缝头处剪开细小的切口。

11

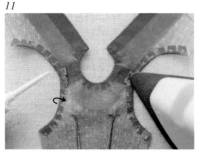

用熨斗把缝头处折过来，然后用布胶临时固定。

12

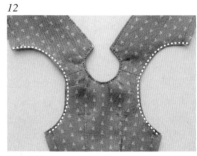

缝合袖筒。

13

将大身前片和大身后片的反面对齐，缝合两侧。

14

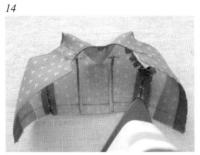

用熨斗把两侧缝头处的布料分开。

15

用熨斗把裙摆的缝头处翻折过来。
→下摆的细节调整，请参考 p.96~99

16

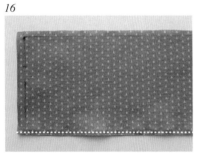

压住裙摆，用缝纫机缝合。

17

在腰部的缝头处，用双股线缝制2.5mm宽的针脚，
用来稍后做褶边。
→褶边请参考 p.100

18

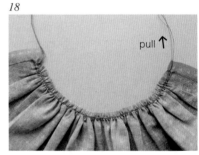

pull ↑

对照大身布料的腰围宽度，调整腰围褶边的长度，
整理均匀后用熨斗熨平。

10. Cut fine slits into the seam allowance of the armholes.   *11.* Fold the seam allowances and apply fabric glue to the seam allowance.
12. Sew the armholes.   *13.* Match the right sides of the front and back bodice sections facing, and sew as pictured.   *14.* Iron open the seam
allowance.   *15.* [refer to p.96−99 for the hem arrangements] Fold the seam allowance of the skirt hem inwards with an iron.
16. Sew the hem.   *17.* Use a machine to sew gathering stitches in the upper seam allowance of skirt. Make the stitch length 2.5mm
and sew two lines on the seam allowance. [refer to p.100 for gathering]   *18.* Gather the fabric to match the width of the bodice waist and iron flat.

19

把大身布料和裙摆布料对齐，裙摆两端的缝头要比大身布料略探出一些。

20

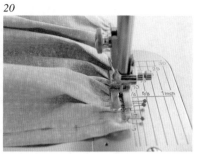

缝合腰部。

21

缝头处向大身布料靠拢，用熨斗熨平。

22

从大身的后开口到开口止处略下一点为止，把后面开口处的缝头斜向内折。

23

压住腰部靠近大身一侧，用缝纫机缝合。

24

缝合以后的状态。

25

后开口对正，从开口止处一直缝合到下摆。

26

用熨斗把两侧缝头处的布料分开。

27

在后边开口处固定按扣，完成。

19. With right sides facing.   20. Sew the waist.
21. Fold the seam allowance to the bodice and iron.   22. Fold the back opening inwards slightly below the stop marker and iron.
23. Sew the waist.   24. Now your stitches are finished.   25. With two sides together and sew.
26. Iron open the seam allowance and turn right side out.   27. Fasten the snaps at the back opening.

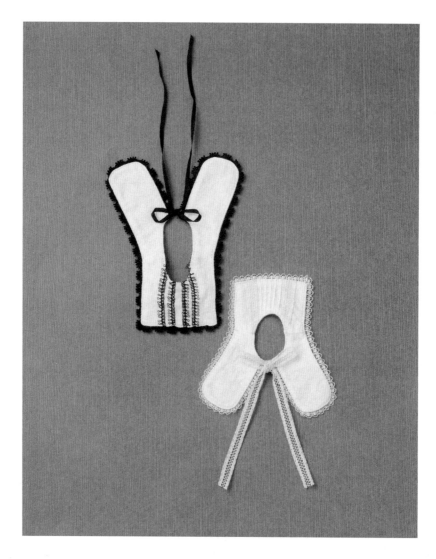

Detachable Collar
活领

与连衣裙和衬衫搭配在一起，能轻松改变整体氛围。需要的布料不多，损耗也很少，建议用这款活领来挑战一下自己的排褶功夫。

锦纶	S	14cm×20cm	蝴蝶结用蕾丝	S,M,L	15cm×2根
	M	14cm×20cm	（5mm宽）		
	L	16cm×20cm	大身用蕾丝	S	10cm
收边用蕾丝	S	30cm	（5mm宽）	M	10cm
（5mm宽）	M	30cm		L	12cm
	L	35cm			

活 领

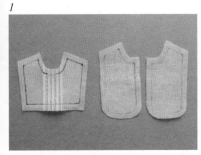

1

配合型纸，裁剪好各部分的布料，用锁边液对布料边缘进行处理。
→排褶的制作方法请参考 p.82、83

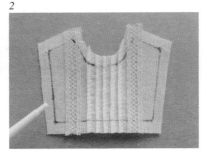

2

用布胶临时固定蕾丝。

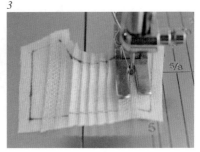

3

缝合蕾丝。

4

排褶嵌入缝头内侧，压住，然后用缝纫机缝合。

5

把大身前片和大身后片的反面对齐，缝合肩部。

6

用熨斗把缝头处的布料分开熨平。

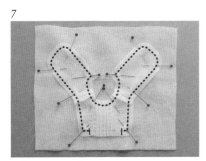

7

准备一片 12cm×12cm 大小的毛裁内衬料，反面对齐。留出折返口后缝合。

8

比照正面布料的边线，把内衬料裁剪好。

9

沿着领口和弧线处的缝头，剪开细小的切口，剪掉边角料。

Just by matching it to a dress or blouse.
It is recommended for the challenge of pin tucks.

1. Arrange the paper templates on the fabric and cut all the sections, then apply fray stopper liquid to all the edges. [refer to p.82–83 for the pin tuck arrangement] 2. Place laces with fabric glue. 3. Sew the laces. 4. Sew the seam allowance of the pin tuck. 5. Match the right sides of the front and back facing, and sew the shoulders. 6. Iron open the seam allowance. 7. Cut the dough about 12cm square for lining. Match the bodice and the lining. Leaving the turn opening and sew as pictured. 8. Cut the lining as pictured. 9. Cut the corners of the seam allowance, cutting fine slits where the fabric curves.

活 领

10

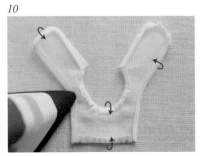

用熨斗把缝头翻折过来。

11

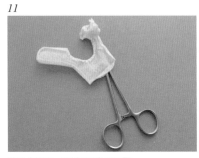

使用小镊子，从返回口翻到正面。

12

用熨斗把边角和弧线处熨平。

13

将折返口的缝头向内翻折，用布胶临时固定。

14

用布胶把蕾丝临时固定在内侧。

15

同样用布胶临时固定用来做蝴蝶结的蕾丝（15cm×2 根）。

16

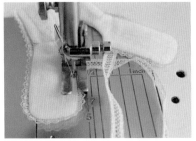

压住领口周围，用缝纫机缝合。

17

缝合折返口。

18

用熨斗熨平，完成。

10. Fold the seam allowance with an iron.　11. Turn the right side out using tailor's awl.　12. Iron into shape.
13. Fold the seam allowance of the turn opening with fabric glue.　14. Place lace on the back side hem with fabric glue.
15. Place the 15cm ribbon on each front opening with fabric glue.　16. Sew the edges.　17. The turn opening is closed.　18. Iron into shape.

Tiered Skirt

荷叶裙

把两片长方形的布料重叠在一起做成荷叶裙。在做好褶边以后，虽然需要用熨斗或临时固定的步骤，
但成品效果非常美观。

锦纶	S	15cm×40cm	皮筋	S,M,L	15cm
	M	17cm×60cm	（3mm 宽）		
	L	20cm×60cm			

荷叶裙

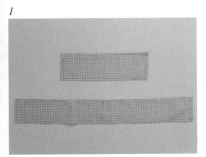

1

配合型纸，裁剪好各部分的布料，用锁边液对布料边缘进行处理。

2

用熨斗把裙子下面（荷叶边）的下摆缝头向内折过来。

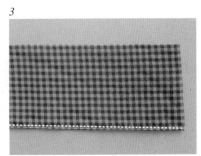

3

压住下摆，用缝纫机缝合。

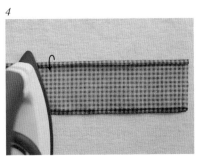

4

用熨斗把荷叶边上面的缝头熨折过来。

5

在腰部的缝头处，用单股线缝制2.5mm宽的针脚，用来稍后做褶边。
→褶边请参考 p.100

6

抽拉↗

对照裙子上半部分下摆的宽度，调整荷叶边的长度。

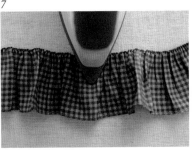

7

整理均匀后用熨斗熨平。

8

在裙子上褶边的正面涂一层薄薄的布胶。

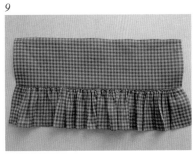

9

把荷叶边放在裙子上半部分的下摆处,暂时固定。

This is a tiered skirt with two rectangular patterns.
After finishing the gathering, the irons and temporary fixings will make the work beautiful.

1. Arrange the paper templates on the fabric and cut all the sections, then apply fray stopper liquid to all the edges.
2. Fold the seam allowance of the frill hem with an iron.   3. Sew the edge.   4. Fold the seam allowance of the frill upper with an iron.
5. Make the stitch length 2.5mm and sew one lines on the seam allowance. [refer to p.100 for gathering]
6. Gather to match the width of fit the skirt hem.   7. Iron flat.   8. Put fabric glue on the hem.   9. Place the frill on the hem.

10

缝合荷叶边。

11

缝合以后的状态。

12

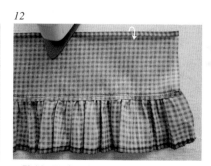

用熨斗把腰围部分向内折熨 3 次。

13

折熨 3 次以后压住，用缝纫机缝合。

14

利用穿皮筋的小工具穿上皮筋。

15

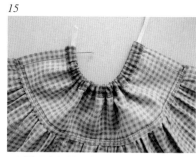

用别针固定好单侧的皮筋，按照 S7cm、M8cm、L8cm 的尺寸收紧裙子的腰围。

16

反面对齐，缝合。

17

用熨斗把缝头处的布料分开熨平。

18

翻到正面，完成。

10. Sew the frill.　*11.* The frill is attached.　*12.* Fold the waist seam allowance 2 times with an iron.
13. Sew the folded.　*14.* String elastic through the waist.　*15.* Gather the waist to [S:7cm　M:8cm　L:8cm] to the finishing line.
16. Sew the back opening together.　*17.* Iron open the seam allowance.　*18.* Turn right side out.

Apron Dress

连衣围裙

有前襟的连衣围裙，可以与衬衫和连衣裙搭配在一起。
选用质地轻薄的锦纶，添加排褶或抽纱刺绣。

锦纶	S	12cm×44cm	锦纶	S	10cm×8cm
（大身和裙摆）	M	14cm×45cm	（内衬）	M	11cm×8cm
	L	15cm×49cm		L	14cm×8cm
锦纶	S	3cm×13cm	按扣	S，M，L 1组	
（腰带）	M	4cm×15cm			
	L	4cm×15cm			

1

配合型纸，裁剪好各部分的布料，用锁边液对布料边缘进行处理。
→大身的细节调整，请参考 p.79~83

2

将大身的正面和反面对齐。

3

上下留出折返口，如图缝合。

4

沿弧线处的缝头剪开细小的切口。小心，不要剪断缝头处的针脚。

5

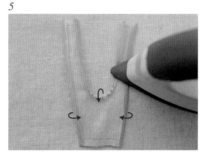

用熨斗把缝头向内侧折好熨平。

6

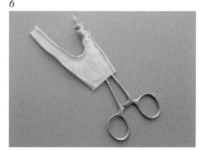

用镊子翻回正面。

7

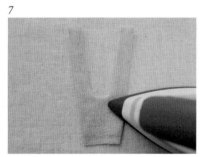

用熨斗熨平。

8

按压好，然后用缝纫机缝合。

9

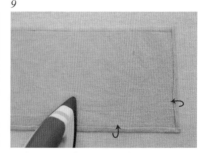

用熨斗，把围裙的下摆和两端的缝头熨平。
→下摆的细节调整，请参考 p.96~99

The Apron dress can be layered on a blouse or dress.
If you put pin tucks or drawn work with a thin cotton loan, it will look great on a layering and it is wonderful.

1. Arrange the paper templates on the fabric and cut all the sections, then apply fray stopper liquid to all the edges.
[refer to p.79−83 for the front arrangements] *2.* Match the right sides of the bodice and lining. *3.* Sew as pictured.
4. Snip the seam allowance. Be careful not to cut the stitches. *5.* Fold and iron the seam allowance. *6.* Turn right side out using a tailor's awl and iron. *7.* Iron to shape. *8.* Sew the edges. *9.* [refer to p.96−99 for the hem arrangements] Fold the seam allowance of the hem and both sides with an iron.

10

压住，用缝纫机缝合。

11

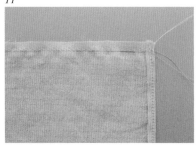

在腰部的缝头处，用双股线缝制 2.5mm 宽的针脚，用来稍后做褶边。
→褶边请参考 p.100

12

pull

抽拉上面的双股线，让褶边聚拢。

13

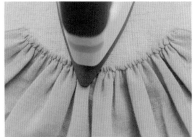

对照腰带的长度调整褶边，整理均匀后用熨斗熨平。

14

将大身和腰带反面对齐。腰带两端的缝头略探出裙摆一些。

15

缝合腰部。

16

把缝头翻到腰带一侧，然后用熨斗熨平。

17

腰带两端的缝头也需要用熨斗翻折过来。

18

腰带上端的缝头也需要用熨斗翻折过来。

10. Sew the edges.　*11.* Use a machine to sew gathering stitches in the upper seam allowance of skirt. [refer to p.100 for gathering] Make the stitch length 2.5mm and sew two lines on the seam allowance.　*12.* Gather the fabric to match the width of the bodice waist.
13. Iron flat.　*14.* With right sides facing, sew the waist belt and the skirt waist together.　*15.* Sew the waist.
16. Fold the seam allowance toward the waist belt using an iron.　*17.* Fold the seam allowance of the both sides.
18. Fold the upper seam allowance of the waist belt.

19

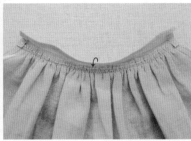

在裙摆的褶边部分涂上布胶，翻过来一半腰带，
临时固定。

20

从正面进行整理，确保美观的效果。

21

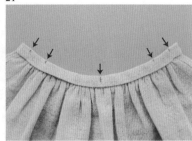

在腰带的正中间、大身后片两端缝合的位置上做
标记。

22

用布胶临时固定大身。

23

压住腰带，用缝纫机缝合。

24

本书中在腰带周围勾勒了一圈边线。

25

把按扣固定在腰带上，完成。

19. Fold the waist belt in half with fabric glue.　*20.* Please check it right side.
21. Mark in the waist belt as pictured.　*22.* Create the bodice to the waist belt with fabric glue.
23. Sew the waist belt.　*24.* The edge stitches are finished now.　*25.* Add snaps to the back waist belt to complete.

Knickerbockers

灯笼裤

裤腿收紧、轮廓蓬松的六分裤。
没有褶边的款式清爽帅气，有褶边的款式可爱别致。

亚麻 / 锦纶	S 18cm×30cm	按扣	S,M,L 1组
	M 20cm×30cm		
	L 23cm×30cm		

灯笼裤

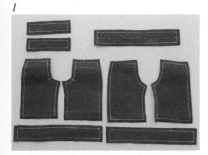

1

配合型纸，裁剪好各部分的布料，用锁边液对布料边缘进行处理。

2

把裤子前裆上方的布料反过来对齐，缝合。

3

沿着弧线的缝头剪开细小的切口。小心，不要剪断缝头处的针脚。

4

用熨斗把缝头处的布料分开熨平。

5

把前裆和后裆的右侧翻过来对齐，缝合。

6

用熨斗把缝头处的布料分开熨平。

7

腰带翻过来对齐，缝合。

8

把缝头翻到腰带一侧，然后用熨斗熨平。

9

用熨斗把腰带的缝头翻折过来。

Plummeted silhouette pants with a squeezed hem.
If there is no frills, it will be neat and pretty style.

1. Arrange the paper templates on the fabric and cut all the sections, then apply fray stopper liquid to all the edges.
2. With the right sides of the left and right front section facing, sew the front rise.   *3.* Cut slits in the seam allowances where the fabric curves.   *4.* Iron open the seam allowance.   *5.* With the right sides of the front and back facing, sew the sides.   *6.* Iron open the seam allowance.   *7.* With the right sides of the trousers and waist belt.   *8.* Fold the seam allowance to the waist belt with an iron.
9. Fold the seam allowance of the waist belt with an iron.

10

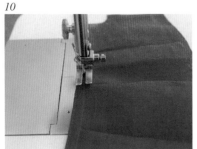

压住腰带，用缝纫机缝合。

11

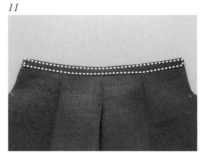

用缝纫机缝合的位置。

12

在裤腿下摆的缝头处，用单股线缝制2.5mm宽的针脚，用来稍后做褶边。
→褶边请参考p.100

13

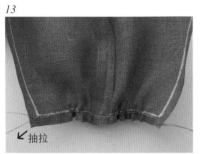

↙抽拉

对照裤腿下摆的宽度，聚拢褶边。

14

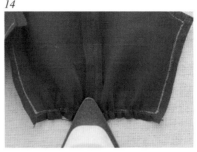

褶边整理均匀以后，用熨斗熨平。

15

裤腿反面对齐，缝合。

16

用同样方法，缝合另一条裤腿。

17

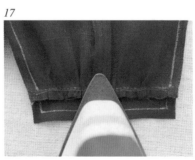

缝头压向裤腿一侧，用熨斗熨平。

18

用熨斗把裤腿下部的缝头翻折过来。

10. Sew the waist belt.   *11*. The waist belt is finished to sew reinforced stitches.
12. Use a machine to sew one line of gathering stitch length 2.5mm on the seam allowance. [refer to p.100 for gathering]
13. Gather to match the width of fit the trousers cuffs.   *14*. Iron flat.   *15*. With right sides of the cuffs and trousers.
16. Sew the cuffs.   *17*. Fold the seam allowance toward to cuffs with an iron.   *18*. Fold the seam allowance of the under cuffs with an iron.

19

用熨斗把褶皱部分下摆的缝头翻过来。

20

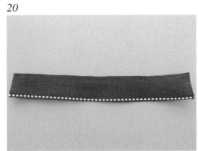

压好,用缝纫机缝合。

21

在上方的缝头处,用单股线缝制2.5mm宽的针脚,用来稍后做褶边。
→褶边请参考 p.100

22

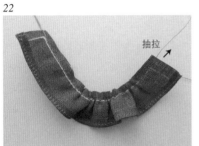

抽拉

对照裤腿的宽度,聚拢褶边。

23

褶边整理均匀以后,用熨斗熨平。

24

用布胶固定在第18个步骤中折好的裤腿下摆缝头处。

25

临时固定褶边。

26

翻到正面来,确认临时固定以后的效果是否美观。

27

压住裤腿的位置,用缝纫机缝合。

19. Fold the seam allowance of the frill hem with an iron.　*20*. Sew the hem.
21. Use a machine to sew one line of gathering stitch length 2.5mm on the upper seam allowance.　*22*. Gather to match the width of fit the cuffs.
23. Iron flat.　*24*. Put fabric glue on the seam allowance of the cuffs hem.　*25*. Attach the frills on the cuffs.
26. Make sure the frills is fine place.　*27*. Sew the cuffs.

28

缝好以后的状态。

29

把从裤子后片的臀部位置开始到后开口标记位置为止对齐，然后缝合。

30

沿着弧线部位的缝头剪开细小的切口。小心，不要剪断缝头处的针脚。

31

用熨斗把缝头处的布料分开熨平。

32

用熨斗把后开口的缝头翻折过来。

33

压住后开口，用缝纫机缝合。

34

把裆下部分反过来对齐，缝合。

35

在裆下的缝头处剪开一个小口。

36

翻到正面，固定好按扣以后完成。

28. The cuffs are finished stitches. 29. With the right sides of the back facing, sew the back rise to the opening marker. 30. Cut slits in the seam allowances where the fabric curves. 31. Iron open the seam allowance. 32. Fold the seam allowance of the back opening with an iron. 33. Sew the back opening. 34. Sew the inseams together. 35. Cut slits into the seam allowance. 36. Turn the right side out. Add snaps to the back opening.

Jacket

夹克衫

宽松的灯笼袖和圆润的娃娃领，使制作出的夹克衫独具特色。
无论是飘逸的蝴蝶结风格，还是对襟的装饰扣，都会让人眼前一亮。

亚麻／天鹅绒	S	18cm×45cm	＜前开襟蝴蝶结款式＞
	M	20cm×50cm	丝带　　S，M，L　10cm×2根
	L	22cm×55cm	（3.5mm宽）
褶皱丝带	S	14cm	＜前开襟按扣款式＞
（领子）	M	15cm	按扣　　S，M，L　3组
	L	16cm	

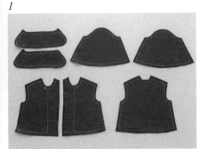

1

配合型纸，裁剪好各部分的布料，用锁边液对布料边缘进行处理。

2

把两片领子的布料翻过来重叠在一起，在外侧缝上装饰线。

3

沿着弧线部位的缝头剪开细小的切口。小心，不要剪断缝头处的针脚。

4

翻到正面，用锥子等调整边角和弧线的线条。

5

用布胶临时固定褶皱丝带。

6

从正面缝合褶皱丝带。

7

缝好以后的状态。

8

将大身前片和大身后片的反面对齐，缝合肩部。

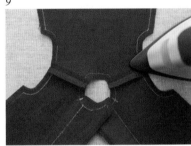

9

用熨斗把缝头处的布料分开熨平。

This jacket has a loose puff sleeve and a rounded collar. It is fastened with a ribbon loosely or it is good to close the front opening with a snap and attach a decorative buttons.

1. Arrange the paper templates on the fabric and cut all the sections, then apply fray stopper liquid to all the edges.
2. Take the collar pieces and match the edges, sew.   *3.* Cut slits in the seam allowance on the round.
4. Turn the piece the right side out, using tailor's awl to neatly push out the corners and curves. Iron to shape.
5. Temporarily fix the frill ribbon with fabric glue.   *6.* Sew the edges.   *7.* The frill ribbon is now attached to the collar.
8. Match the front and back of the bodice by the shoulders and sew.   *9.* Iron open the seam allowance.

10

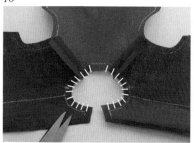

沿着领子缝头剪开细小的切口。

11

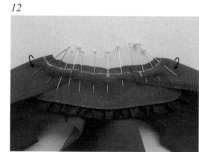

用别针固定领子和大身。

12

翻到反面，用别针固定。

13

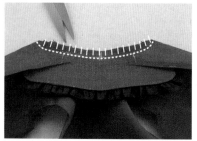

缝上装饰线，在缝头处剪开细小的切口，剪掉边角料。

14

翻回到正面，用熨斗熨平。

15

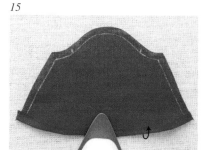

用熨斗把袖口的缝头折过来。

16

用单股线缝制2.5mm宽的针脚，用来稍后做褶边。
→褶边请参考 p.100

17

把褶边聚拢到装饰线内侧（S: 4cm M: 4.5cm L: 4.5cm）。

18

把褶边调整均匀以后，熨平。然后压住，用缝纫机缝合。

10. Cut slit in the seam allowance of the neckline.   *11*. Pin with right sides facing.   *12*. Fold the front facing so it faces inwards.
13. Sew the neckline. Cut fine slits in the seam allowance of the neckline.   *14*. Turn the front facing the right side out, iron to shape.
15. Fold the seam allowance of the sleeve opening inward and iron.   *16*. Machine sew one line of gathering stitches 2.5mm in length in the seam allowance.   [refer to p.100 for gathering]   *17*. Gather to [S:4cm M:4.5cm L:4.5cm] with the finished line.   *18*. Iron flat.

19

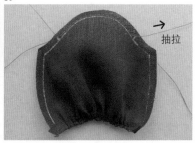

→ 抽拉

从肩膀处缝头的标志到标志之间，用单股线缝制2.5mm宽的针脚，用来稍后做褶边。

20

比照大身的袖筒长度，聚拢褶边，把线打结。然后用熨斗熨烫平整。

21

将袖子与大身反面对齐，缝合。

22

肩膀处的缝头与袖筒的缝头要一点一点地对齐，用缝纫机压平，反复而缓慢地仔细缝合。

23

把袖子缝在大身上。缝头向袖子一侧按平，用熨斗熨烫平整。

24

将大身前片与后片反面对齐，沿着袖口、两侧、下摆缝合。

25

在两侧的缝头处剪开小口。

26

翻回正面，用熨斗把缝头处的布料分开熨平。

27

将两边下摆的背面翻出来。如图，用别针临时固定。

19. Machine sew one line of gathering stitches 2mm in length in the sleeve cap seam allowance from marker to marker.
20. Gather the shoulders until the width fits the armhole and iron.   *21*. Match the side edge of the sleeve to the bodice and gradually, sew the shoulder of the sleeve to armhole.   *22*. Raise the machine presser foot a number of time while sewing to gradually align the seam allowances of the sleeve caps and armholes.   *23*. Now the sleeves are attached. Place the seam allowance toward the sleeves and iron.
24. With the right sides of the front and back bodice facing, sew them together.   *25*. Cut slits into the seam allowance of the pits.
26. Turn right side out. Iron open the seam allowance.   *27*. Pin the hem of the front facing.

28

缝合下摆。

29

剪掉缝头的边角料。

30

缝合完成以后的状态。

翻回正面，用熨斗熨烫平整。用熨斗把下摆的缝头折好。

31

从领口开始，一边压住一边缝合。

32

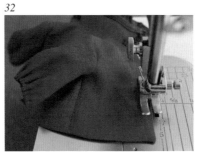

顺着前开襟、下摆、前开襟、领口一周的顺序，仔细缝合。

33

缝合完成以后的状态。

34

在领口固定一条10cm的丝带。（也可以在前开襟缝合按扣）

35

用单股刺绣线刺绣一枚三叶草。

36

完成。

28. Sew the hem of the front facing.   29. Cut the corners.   30. Turn right side out of the front facing and iron. Fold the hem.
31–33. Sew around the edges from neckline along the front opening and the hem, then back up to the neckline.
34. Sew the 10cm ribbon to the front opening. Or you can add snaps to the front opening.
35. Take a single embroidery and sew reverse stitches for initial.   36. Complete the jacket.

Coat
外套

A 字版百搭外套。领子的型纸与夹克衫的型纸通用。

立起来可以体现撞色风格，放下去可以变身可爱小圆领。同一款服饰能享受不同的乐趣。

羊毛 / 亚麻	S	15cm×48cm	按扣	S,M,L　4组
	M	17cm×55cm		
	L	22cm×60cm		
褶皱丝带	S	14cm		
（领子）	M	15cm		
	L	16cm		

外套

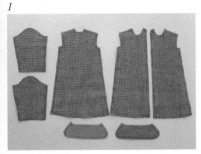

配合型纸，裁剪好各部分的布料，用锁边液对布料边缘进行处理。

把两片领子的布料翻过来重叠在一起，在外侧缝上装饰线。

在弧线部位的缝头处剪开细小的切口。小心，不要剪断缝头处的针脚。

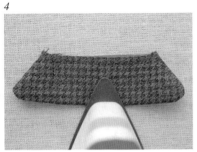

翻到正面，用锥子或镊子调整边角和弧线的线条。

压齐，缝合。

缝合完毕。

将大身前片和大身后片的反面对齐，缝合肩部。

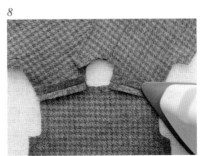

用熨斗把缝头处的布料分开熨平。

沿着领子缝头处剪开细小的切口。

The collar pattern is common with the jacket, When opened, it turns into a tailored collar, and when closed, it turns into a cute round collar. Enjoy the difference.

1. Arrange the paper templates on the fabric and cut all the sections, then apply fray stopper liquid to all the edges.
2. Take the collar pieces and match the edges, sew.   3. Cut slits in the seam allowance on the round.
4. Turn the piece the right side out, using tailor's awl to neatly push out the corners and curves. Iron to shape. 5. Sew the edges of the collar.
6. Finished to sew the edges.   7. Match the front and back of the bodice by the shoulders and sew.
8. Iron open the seam allowance.   9. Cut slits in the seam allowance of the neckline.

10

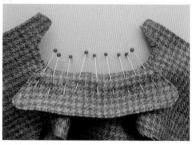

用别针临时固定领子和大身。

11

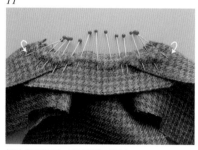

折边翻到反面，用别针临时固定。

12

缝制装饰线，再沿着缝头剪开细小的切口，剪掉边角料。

13

翻到正面，用熨斗熨烫平整。

14

用熨斗把袖口的缝头折好。

15

缝合。

16

从肩膀处缝头的标志到标志之间，用单股线缝制2mm 宽的针脚，用来稍后做褶边。

17

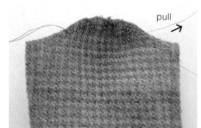

比照大身的袖筒长度，聚拢褶边，把线打结。

18

整理肩膀处的褶边，然后用熨斗熨烫平整。

10. Pin with right sides facing. *11.* Fold the front facing so it faces inwards. *12.* Sew the neckline. Cut fine slits in the seam allowance of the neckline. *13.* Turn the front facing the right side out, iron to shape. *14.* Fold the seam allowance of the sleeve opening inward and iron. *15.* Sew the sleeve opening. *16.* Machine sew one line of gathering stitches 2.5mm in length in the seam allowance. *17.* Gather the shoulders until the width fits the armhole. *18.* Iron flat.

外套

19

将袖子与大身反面对齐，缝合。

20

肩膀处的缝头与袖筒的缝头要一点一点地对齐，用缝纫机压平，反复而缓慢地仔细缝合。

21

慢慢向前推进，实现美观的缝合效果。

22

把袖子缝在大身上。

23

缝头向袖子一侧按平，用熨斗熨烫平整。将大身前片与后片反面对齐。

24

沿着袖口、两侧、下摆缝合。

25

在两侧的缝头处剪开小口。

26

翻回正面，用熨斗把缝头处的布料分开熨平。

27

两边下摆的背面翻出来。如图，用别针临时固定。

19–22. Match the side edge of the sleeve to the bodice and gradually, sew the shoulder of the sleeve to armhole. Raise the machine presser foot a number of times while sewing to gradually align the seam allowances of the sleeve caps and armholes.
23. Place the seam allowance toward the sleeves and iron. With the right sides of the front and back bodice facing.
24. Sew them together. *25.* Cut slits into the seam allowance of the pits.
26. Turn right side out. Iron open the seam allowance. *27.* Pin the hem of the front facing.

28

缝合下摆，剪掉缝头的边角料。

29

翻回正面，用熨斗把下摆的缝头折好。

30

从领口开始，一边压住一边缝合。

31

从领口过渡到前襟。

32

从前襟过渡到下摆。

33

按照领口、前襟、下摆、前开襟、领口一周的顺序，仔细缝合。缝合完成以后的状态。

34

用熨斗整理领子的形状。

35

固定好按扣和小珠子。

36

这种领子的形状，即使不做撞色款也很可爱。

28. Sew the hem. Cut the corners.   29. Turn right side out of the front facing and iron. Fold the hem.
30–33. Sew around the edges from neckline along the front opening and the hem, then back up to the neckline.
34. Iron to shape.   35. Add snaps and beads.   36. It is also cute not to make the shape of the collar tailored.

Hat
帽子

分为通用的 1/6 娃娃用以及小布娃娃（Blythe）等大头娃娃用的帽子。
可以在丝带款和花饰款中尽情选择。

亚麻　　1/6 娃娃尺寸　　12cm×30cm
　　　　小布娃娃尺寸　　35cm×50cm

帽子

1

配合型纸，裁剪好各部分的布料，用锁边液对
布料边缘进行处理。
→小布娃娃的尺寸请参见 p.72

2

布料侧边反面对齐。

3

缝合

4

用熨斗把缝头处的布料分开熨平。

5

将帽顶布料和侧边布料反面对齐。

6

如果难以用缝纫机缝合，可以使用双股线倒针
缝合。

7

缝合完毕。

8

在缝头处剪开细小的切口。小心，不要剪断缝头
处的针脚。

9

翻回到正面。

A hat for the average 1/6 doll and a doll with a big head such as Blythe.
Please enjoy wrapping ribbons and decorating flowers.

1. Arrange the paper templates on the fabric and cut all the sections, then apply fray stopper liquid to all the edges. [Blythe size p.72]
2. Match the side crown together inside out.   *3.* Sew the side crown.   *4.* Iron open.   *5.* Match the top crown and side crown together inside out.   *6–7.* If it is difficult to sew with sewing machine, you can sew by hand with two threads to back stitch.
8. Cut slits the seam allowance. Be careful not to cut the stitches.   *9.* Turn right side out.

10

将帽檐反面对齐。

11

缝合。

12

用熨斗把缝头处的布料分开熨平。

13

缝好以后的状态。

14

将反面对齐，在外侧缝合。

15

沿着缝头剪开细小的切口。小心，不要剪断缝头处的针脚。

16

翻到正面，用熨斗熨烫出平整而美观的圆形。整理好。

17

缝合。

18

用别针把帽檐和帽体反面对齐，临时固定。

10. Match the brim together inside out.　*11*. Sew together.　*12*. Iron open.　*13*. Sew two pieces.
14. Match the brim inside out and sew.　*15*. Cut slits the seam allowance.　*16*. Turn right side out. Iron to shape.
17. Sew the edges.　*18*. Pin the brim and crown inside out together.

帽子

19

如果难以用缝纫机缝合，可以使用双股线倒针缝合。

20

缝好以后的状态。

21

沿着缝头处剪出细小的切口。

22

把缝头压向帽体一侧，用熨斗熨烫平整，完成。

23

搭配丝带的款式。

24

多款丝带堆叠在一起，做成花饰的款式。

25

搭配细丝带和蝴蝶结的款式。

19−21. If it is difficult to sew with sewing machine, you can sew by hand with two threads to back stitch.
22. Fold the seam allowance toward to the crown with an iron.   *23.* This is arrangement with laces.
24. Add the several laces are together with running stitch.   *25.* This is arrangement with thin ribbon.

帽子

1

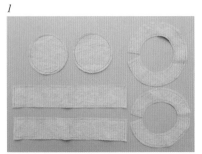

小布娃娃的帽子需要搭配内衬。与正面相同，配合型纸，裁剪好各部分的布料，用锁边液进行处理。

2

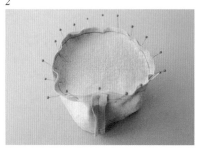

侧面布料反面连成一个环形，把缝头分开。像正面一样，把各部分的布料放在一起对齐，用别针临时固定。

3

缝合侧面和帽顶的布料，将侧面和帽顶反面对齐，用别针临时固定。

4

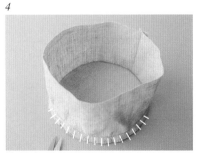

沿着缝头处剪出细小的切口。小心，不要剪断缝头处的针脚。

5

用熨斗把缝头分开。用同样方式制作内衬。

6

外布料和内衬内侧相扣，叠放到一起。

7

按照前款第10~17的顺序缝合，把布料反面对齐，用别针临时固定。

8

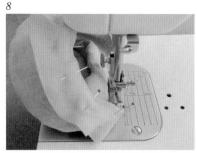

把帽顶和侧面缝合。

9

沿着缝头处剪出细小的切口，用熨斗把缝头熨烫到帽体一侧，整理平整，完成。

Huge Hat

1. Arrange the paper templates on the fabric and cut all the sections, then apply fray stopper liquid to all the edges. The huge hat needs back fabric.
2. Match the side crown together inside out and sew. Iron open. Match the top crown and side crown together inside out.
3. Sew together. *4.* Cut slits the seam allowance. Be careful not to cut the stitches. *5.* Iron open. Make the back fabric in the same way.
6. Set the back fabric inside the crown of the right side. *7.* Sew the brim. [Refer to 10–17.] Pin the crown and the brim together. *8.* Sew together.
9. Cut slits the seam allowance. Turn right side out. Fold the seam allowance toward to the crown with an iron.

Bag

手袋

做手袋的时候，可以使用无须担心端口开裂的皮料，绒面皮及合成皮。
建议选用质地轻薄的材料。如果担心包体强度，可以在连接处用线缝合固定。

合成皮　8cm×10cm

刺绣线　根据个人喜好

手袋

1

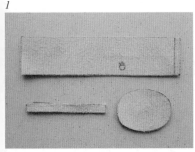

参照水桶包的型纸，剪好各个部分的材料。

2

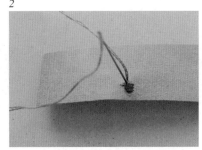

绣上自己喜欢的图案。此处需要用双股绣线。

3

本款手袋绣了一颗草莓。

4

在接头处涂抹皮料用黏合胶。

5

两端重合在一起，完成水桶包包体。

6

在包底部分的边缘涂抹木工胶。

7

把包体和包底粘贴在一起，固定。

8

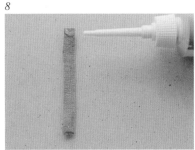

在提手部分的两端涂抹木工胶。

9

固定提手，完成。

For the bag, use leather, suede, or synthetic leather. It is recommended to use thin materials.
If the strength is insecure, you can sew the edges.

1. Arrange the paper templates on the fabric and cut all the sections.   *2.* Use two embroidery threads.   *3.* The strawberry embroidery is finished.
4. Put leather glue on the edges.   *5.* Match the edges together and make a tube.   *6.* Put leather glue on the edges.
7. Match together.   *8.* Put leather glue on the edges of the handle.   *9.* Paste together.

手袋

1

参照托特包的型纸，裁剪好各个部分的材料。

2

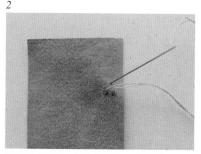

绣上喜爱的图案。此处需要使用双股绣线。

3

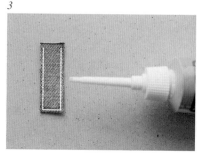

在包体侧面的材料上涂抹皮革黏合胶。

4

包体完成以后，粘贴固定包底部位的材料。然后在包底和侧面的材料上折叠出印痕。

5

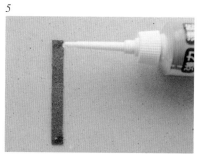

在提手部分的两端涂抹木工胶。

6

粘贴固定提手的部分，完成。

7

可以调整提手部分的长短和包体的大小，款式多样，妙趣横生。

1. Arrange the paper templates on the fabric and cut all the sections.   *2*. Use two embroidery threads.   *3*. Put leather glue on the edges.
4. Fold the bag and paste together with side pieces. And fold the bottom and the sides.   *5*. Put leather glue on the edges.   *6*. Paste together.
7. You can make various arrangements to change the length of the handle or the size of the bag.

变化
Variation

大身部分调整

领子部分调整

袖子部分调整

下摆部分调整

大身部分调整

a.
Lace Collage
蕾丝拼接

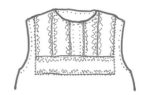

大身前片的调整当中，最简单的就是领口部位。只要把蕾丝摆在上面缝合即可。

1

按个人喜好选配蕾丝，摆放在大身前片的中央。

2

侧面也同样摆放蕾丝。

3

左右镜像，使蕾丝相互搭接一点儿。

4
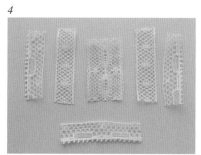
把蕾丝从大身前片上拿下来。

5

用布胶把蕾丝一枚一枚临时固定好。

6

缝合。

7

缝好以后的样子。

8

依照大身的尺寸裁剪蕾丝。

9

多缝几条纵向的蕾丝，可以体现出可爱的风格。

1. Place your favorite size lace in the center of the front bodice. *2.* Place the lace in the both of sides.
3. Place the laces little by little symmetrically. *4.* Take off the laces from the front bodice.
5. Apply fabric glue the laces. *6–7.* Sew the laces. *8.* Cut the laces. *9.* It is cute just sewing several laces vertically.

大身部分调整

b.
Lace Frill
蕾丝褶边

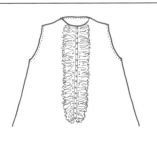

如果蕾丝有剩余，可以捏在一起做褶边。

1

根据个人喜好，在大身前片的中央画一条直线。

2

准备一根长度为直线 2 倍的蕾丝，在蕾丝中间画一条线。

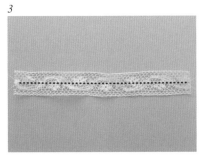

3

每一个针脚间距 2mm 左右,缝一根单股的褶边线。
→褶边请参考 p.100

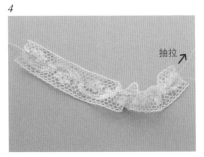

4

抽拉↗

依照大身前片上的线长聚拢褶边,将线打结。

5

整理褶边,用熨斗熨烫。

6

在大身前片的线上涂抹布胶。

7

临时固定蕾丝。

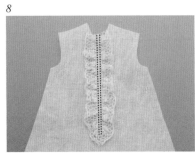

8

在针脚上面放一根丝带,临时固定好以后缝合。

9

作为装饰,可以搭配一些小珠子或小扣子。

1. Draw the line of desired length in the center of the front bodice.   *2.* Prepare a lace that is about twice as long as the line.
Draw the line in the center of the lace.   *3.* Machine sew one line of gathering stitches 2mm in length in the center. [refer to p.100 for gathering]
4. Gather the lace until the width fits the line.   *5.* Iron flat.   *6.* Apply fabric glue on the line.   *7.* Temporarily the lace.
8. Apply fabric glue on the center of the lace. Place the thin ribbon and sew.   *9.* It is cute for decoration with beads or buttons for finishing.

大身部分调整

c.
Lace Pleats
蕾丝褶皱

单侧蕾丝可以使褶皱体现出华丽的风格。如果中间的细丝带选用其他颜色，则更加别致。

1

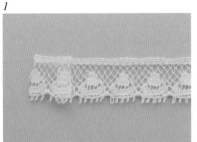

把蕾丝叠出来 5mm 的褶皱。

2

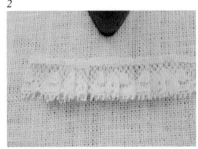

用布胶固定，这样有利于后面的步骤操作。

3

按照需要的宽度制作褶皱。

4

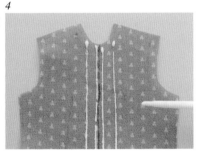

在大身前片上涂抹布胶。

5

临时固定褶皱。

6

在中间临时固定一根细丝带。

7

缝合。

8

缝好以后的状态。

9

完成后可以搭配一些小珠子或小扣子。

1. About 5mm pleats to the lace.　*2.* It is easy to make with fabric glue.　*3.* Make pleats to the width you want to use.
4. Apply fabric glue on the front bodice.　*5.* Temporarily the lace.　*6.* Apply fabric glue on the center of the lace. Place the thin ribbon.
7–8. Sew the ribbon.　*9.* It is cute for decoration with beads or buttons for finishing.

大身部分调整

d.
Pin Tuck

排褶

在活领单品上应用排褶技巧。当然，同样也能在衬衫和连衣围裙上使用。

1

准备一张略大于型纸的布料。

2

用熨斗沿着布料的标志处折叠。

3

根据个人喜好，在折痕处留出一点儿距离，然后画出缝合线（图中距离折痕 1mm）。

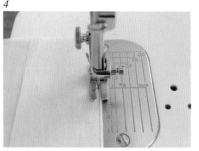

4

在缝合线上缝合。

5

缝好以后的状态。

6

打开布料，用熨斗把刚才缝好的折痕向外侧打开。

7

再沿着刚才的折痕，用熨斗折回去（这次也是根据个人喜好决定宽度）。

8

画缝合线，缝合。

9

打开布料，用熨斗把刚才缝好的折痕向外侧打开。

1. Prepare a fabric that is larger than the pattern you want to use.   *2.* Fold along the weave with an iron.
3. Draw a seam line from the crease at the desired width (the photo is 1 mm from the crease).   *4–5.* Sew the line.
6. Iron open and iron toward to outside.   *7.* Fold in the same way along the texture.   *8.* Draw the line and sew.   *9–10.* Iron open and outward.

10

排褶缝好以后的状态。

11

按照型纸描绘轮廓线。

12

剪裁。

13

用锁边液对布料边缘进行处理。

14

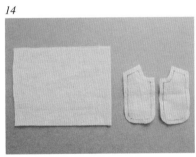

如果用排褶的方式制作活领，需要把排褶位置的布料预留大一些。

15

排褶聚拢到一起。

16

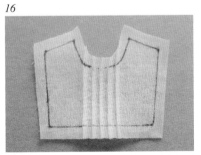

配合型纸，裁剪好各部分的布料，用锁边液对布料边缘进行处理。

17

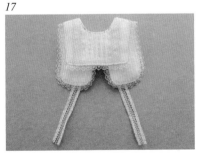

使用活领时，还可以在排褶的边缘放好蕾丝，缝合起来。

11. Trace the pattern on the fabric.   *12*. Cut the section.   *13*. Apply fray stopper liquid to the edges.
14. For detachable collar, Cut the parts you want to pin−tack large.   *15*. Make pin tucks.   *16*. Arrange the paper templates on the fabric and cut the section, then apply fray stopper liquid to all the edges.   *17*. This detachable collar has laces on the sides of the pin tucks.

领 子 部 分 调 整

e.
Ruffled Collar
荷叶领

大气美观、华丽无比的荷叶领。如果选用大幅蕾丝的材料，成品效果更加理想。

1

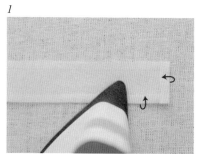

用熨斗折叠褶皱部分的缝头。

2

压好以后缝合。

3

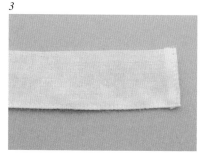

缝合以后的放大图。

4

在缝头处，用单股线 2mm 左右的针脚缝制，用于制作褶皱。

5

褶皱线的位置要略高于缝合线。
→褶边请参考 p.100

6

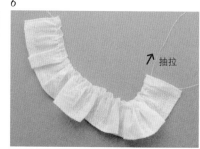

依照大身的领口长度聚拢褶皱，将线打结。

7

用熨斗熨烫，把褶皱调整均匀。

8

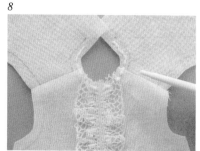

在领口周围的缝头处涂抹布胶。

9

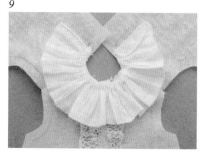

临时固定领子。

1. Fold the seam allowance of the ruffled with an iron.   *2–3.* Sew the edges.
4. Use a machine to sew one line of gathering stitch length 2mm on the upper seam allowance.
5. Sew gathering stitch slightly above the finished line. [refer to p.100 for gathering]   *6.* Gather the fabric to match of the width of the neckline.
7. Neaten the spacing of the gathering and iron flat.   *8.* Apply fabric glue to the seam allowance.   *9.* Place the ruffled collar.

10

在大身后片折边处的缝头上涂抹布胶。

11

折边反面对齐，与领子重叠，临时固定。

12

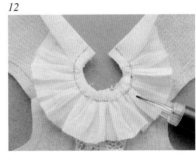

画出缝合线。

13

缝合领口。

14

在缝头处剪出细小的切口，剪掉边角料。

15

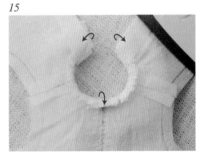

翻到正面，用熨斗把领子的缝头向内侧熨平。

16

沿着后开口、领口、后开口的顺序缝合。

17

缝合以后的状态。

18

在领子上喷水雾，然后用熨斗熨烫，完成。

10. Apply fabric glue to the seam allowance of the back facing.　11. Match the back facing.　12. Trace the finishing line.
13. Sew the neck line.　14. Cut slits in the seam allowances and cut the corners.　15. Turn right side out of the back facing.
Fold inside the seam allowance of the collar.　16–17. Sew the edges as pictured.　18. Iron with little water into shape.

领子部分调整

f.
Peter Pan Collar
圆领

领边的线条柔和，是一款易于缝合的圆领。如果服装材料较厚，建议选用薄料的内衬。

1

在布料上画出一对领子的形状，裁剪的时候要稍微大一圈。 然后剪出另外一片同样大小的布料。

2

将两片反面对齐，缝合外侧的装饰线。

3

留下缝头，把领子剪下来。剪掉边角料。

4

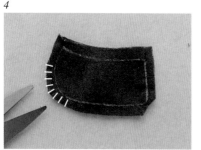

在缝头处剪出细小的切口。小心，不要剪断缝头处的针脚。

5

翻到正面,用锥子或镊子调整边角和弧线的线条。

6

在缝头上固定小毛边。

7

沿着大身领口周围的缝头处剪出细小的切口。

8

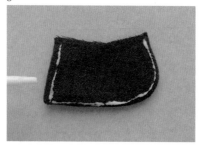

如果要搭配领子上的蕾丝，可以先把布胶涂抹在领子内侧。

9

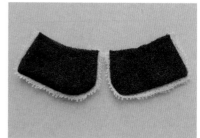

临时固定蕾丝。

1. For the collar take two pieces of the same size and draw the collar on one piece. *2.* Take the collar pieces and match the edge, sew along the outer seam line. *3.* Cut the collar sections out, leaving seam allowance, and cut the corners. *4.* Cut slits in the corner. Be careful not to cut the stitches. *5.* Turn right side out with tailor's awl then iron into shape. *6.* Apply fray stopper liquid to the seam allowance. *7.* Cut slits in the seam allowance. *8.* If you want to put the lace around, then apply fabric glue on the edge of the back collar. *9.* Place the laces.

10

缝合。

11

注意左右对称，临时固定领子。

12

在领子上画好缝合线，在大身后片的折边缝头处涂抹布胶。

13

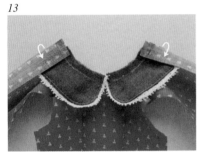

折边反面对齐，与领子重合，临时固定。

14

缝合领子。在领子的缝头处剪出细小的切口，剪掉边角料。

15

折边翻到正面，用熨斗把领子的缝头熨烫到内侧。

16

沿着后开口处、领口一圈，再回到后开口处的顺序缝合。

17

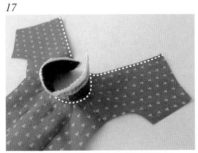

缝合以后的状态。

18

圆领完成。

10. Sew the edges.　11. Make sure to be symmetric, apply fabric glue on the seam allowance of the neck and place the collars.
12. Trace the finishing line and apply fabric glue on the seam allowance of the back facing.　13. Fold the back facing.
14. Sew the collars, cut slits in the seam allowance and corners.　15. Turn right side out of the back facing. Fold inside the seam allowance of the collar.
16–17. Sew the edges as pictured.　18. Iron into shape.

领子部分调整

g.
Lace Stand Collar
蕾丝立领

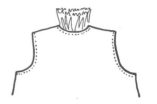

布料的松紧蕾丝，可以帮助您简单地做出美丽的立领。可长可短。

1

准备一条蕾丝（S:11cm M:12cm L:15cm）。

2

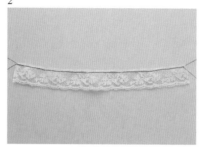

在蕾丝的上部，用单股线缝出一条宽度为2mm
针脚，用于稍后制作褶边。
→褶边请参考 p.100

3

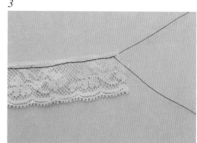

褶边的放大图。

4

抽拉

单侧打结，拉动另外一侧的线，聚拢褶边。

5

依照领口的尺寸给线打结，然后用熨斗熨烫褶边。

6

带着内衬，在大身的领口涂抹布胶。

7

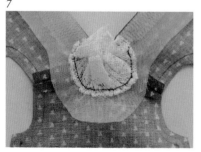

临时固定蕾丝。

8

沿着后开口处、领口一圈，再回到后开口处的顺序，
缝合。

9

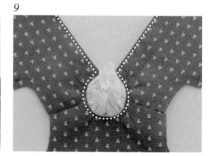

蕾丝领完成。

1. The lace for the collar width is [S:11cm　M:12cm　L:15cm].
2-3. Use a machine to sew one line of gathering stitch length 2mm on the upper seam allowance. [refer to p.100 for gathering]
4. Tie the ends at the one side and pull the bottom thread from one side.　*5*. Match the width of fit the neck line and tie the threads.
6. Apply fabric glue on the seam allowance.　*7*. Place the lace collar.　*8-9*. Sew the edges as pictured.

袖子部分调整

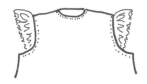

h.
Lace Cap Sleeve
蕾丝鸡翼袖

这款蕾丝袖可以推荐给初次搭配袖子的爱好者。只要左右宽度一致，就能制作出优雅而美丽的成品。

1

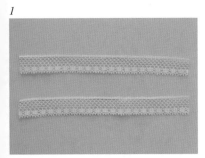

准备两根蕾丝（S：8cm M：9cm L：10cm）。推荐选用宽度为1~1.5cm的蕾丝。

2

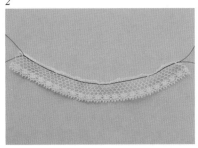

在蕾丝的上部，用单股线缝出一条宽度为2mm针脚，用于稍后制作褶边。

3

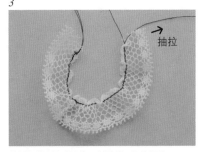

→ 抽拉

单侧打结，拉动另外一侧的线，聚拢褶边。

4

按照一定长度（S：3.5cm M：4cm L：4.5cm），把褶边聚拢到一起，线打结。

5

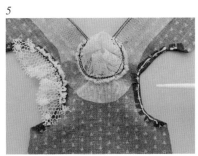

在袖筒的缝头处剪出细小的切口，用布胶临时固定。

6

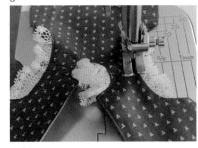

缝合。

7

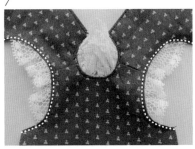

缝合以后的状态。

8

将大身反面对齐，两侧缝合。

9

翻到正面，完成。

1. The lace for the collar width is [S:8cm M:9cm L:10cm]. *2.* Use a machine to sew one line of gathering stitch length 2mm on the upper seam allowance.
3. Tie the ends at the one side and pull the bottom thread from one side.   *4.* Match the width [S:3.5cm M:4cm L:4.5cm] and tie the threads.
5. Cut fine slits into the seam allowance of the armholes. Fold the seam allowances and temporarily the lace with fabric glue.
6-7. Sew the armholes.   *8.* With right sides of front and back facing, sew together.   *9.* Turn right side out.

袖子部分调整

i.
Set-in Sleeve
圆袖

强调轮廓感的长袖。袖子上的小褶边营造可爱的立体感。

1

配合型纸,裁剪好各部分的布料,用锁边液对布料边缘进行处理。

2

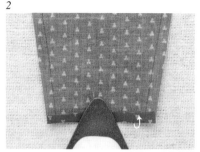

用熨斗把袖口的缝头折过来。

3

如果在袖口上搭配蕾丝,可以使用布胶。

4

临时固定蕾丝。

5

缝合。

6

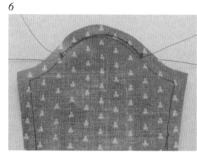

从肩袖的缝头标志处到标志处,缝出2mm宽的针脚,用于制作褶边。

7

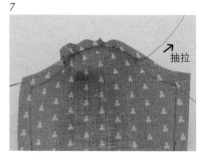

抽拉

依照大身的袖筒长度聚拢褶边,线打结。

8

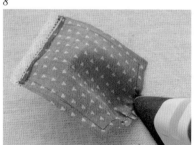

用熨斗熨烫褶边。

9

将大身和袖子反面对齐,缝合。

1. Arrange the paper templates on the fabric and cut all the sections, then apply fray stopper liquid to all the edges.
2. Fold the seam allowance of the sleeve opening.　*3.* If you want to put the lace on the sleeve opening, apply fabric glue.
4. Temporarily the lace with fabric glue.　*5.* Sew the sleeve opening.　*6.* Machine sew one line of gathering stitches 2mm in length in the seam allowance.
7. Gather the shoulders until the width fits the armhole.　*8.* Iron flat.

10

将肩袖的缝头和袖筒缝头仔细对齐，用缝纫机压住，分几次缓慢缝合。

11

缝合速度放慢，可以确保完美的缝合效果。

12

大身和袖子缝到一起了。

13

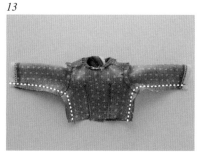

缝头向袖子一侧熨平，大身前侧和大身后侧反面对齐，沿着袖口、两侧、下摆的顺序缝合。

14

在两侧的缝头处剪出小切口。

15

翻回正面，用熨斗把缝头分开。

16

完成。

9–12. Match the side edge of the sleeve to the bodice and gradually, sew the shoulder of the sleeve to armhole.
Raise the machine presser foot a number of times while sewing to gradually align the seam allowances of the sleeve caps and armholes.
13. Place the seam allowance toward the sleeves and iron. With the right sides of the front and back bodice facing. Sew them together.
14. Cut slits into the seam allowance of the pits. 15. Turn right side out. Iron open the seam allowance. 16. Iron into shape.

袖子部分调整

j.
Ballon Sleeve
泡泡袖

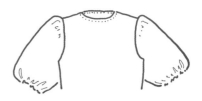

因为肩袖部线条柔和，所以缓慢而仔细地缝纫才能带来完美的效果。

1

配合型纸，裁剪好各部分的布料，用锁边液对布料边缘进行处理。

2

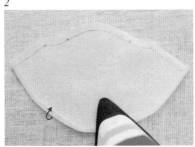

用熨斗把袖口缝头折过来。

3

涂抹布胶。

4

临时固定蕾丝。

5

在缝头处缝上 2mm 的针脚，用于做褶边。

6

在缝头处缝上 2mm 的针脚，用于做褶边。

7

按照一定长度（S: 4cm M: 4cm L: 4.5cm）把褶边聚拢到一起，将线打结。

8

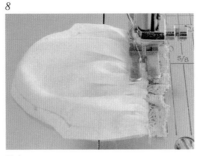

缝合。

9

缝合以后的状态。

1. Arrange the paper templates on the fabric and cut all the sections, then apply fray stopper liquid to all the edges. *2.* Fold the seam allowance of the sleeve opening. *3.* Apply fabric glue. *4.* Temporarily the lace with fabric glue. *5.* Machine sew one line of gathering stitches 2mm in length in the seam allowance. *6.* Tie the ends at the one side and pull the bottom thread from one side. *7.* Gather the width [S:4cm M:4cm L:4.5cm] from finished line to finished line. Tie the threads. *8-9.* Sew the sleeve opening.

10

从肩袖的缝头标志处到标志处，缝出 2mm 宽的针脚，用于制作褶边。

11

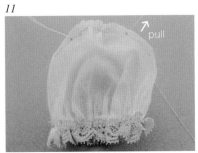

pull

依照大身的袖筒长度聚拢褶边，线打结。

12

用熨斗熨烫褶边。

13

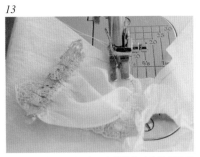

将大身和袖子反面对齐。肩袖的缝头和袖筒缝头仔细对齐，用缝纫机压住，分几次缓慢缝合。

14

袖子缝好了。

15

将大身反面对齐，沿着袖口、两侧、下摆的顺序缝合。

16

缝合完成。

17

在腋下的缝头处剪出小切口。

18

翻回正面，用熨斗把缝头分开，完成。

10. Machine sew one line of gathering stitches 2mm in length in the seam allowance.   11. Gather the shoulders until the width fits the armhole.
12. Iron flat.   13-14. Match the side edge of the sleeve to the bodice and gradually, sew the shoulder of the sleeve to armhole.
Raise the machine presser foot a number of times while sewing to gradually align the seam allowances of the sleeve caps and armholes.
15-16. Place the seam allowance toward the sleeves and iron. With the right sides of the front and back bodice facing. Sew them together.
17. Cut slits into the seam allowance of the pits.   18. Turn right side out. Iron open the seam allowance.

袖子部分调整

k.
Band Cuffs
束带袖

束带袖的体积比泡泡袖小一些，就像是没有褶边的泡泡袖。

1

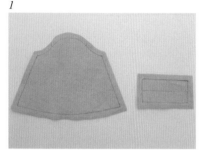

配合型纸，裁剪好各部分的布料，用锁边液对布料边缘进行处理。

2

在袖口的缝头处缝上 2mm 的针脚，用于做褶边。

3

抽拉

单侧打结，抽拉另一侧的线，聚拢褶边。

4

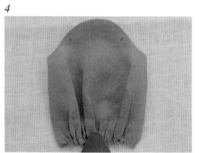

依照束带的长度把褶边聚拢到一起，将线打结。用熨斗熨烫平整。

5

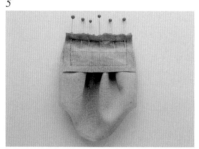

将袖口与束带反面对齐。

6

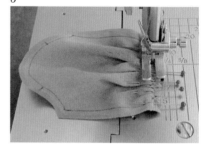

缝合袖口。

7

缝合以后的状态。

8

用熨斗把缝头向束带一侧压倒熨平。

9

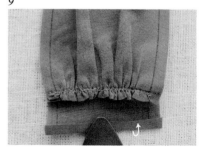

用熨斗把束带的缝头打开熨平。

1. Arrange the paper templates on the fabric and cut all the sections, then apply fray stopper liquid to all the edges.
2. Machine sew one line of gathering stitches 2mm in length in the seam allowance.
3. Gather the sleeve opening until the width fits the cuff.　*4.* Iron flat.　*5.* Match the sleeve and cuff.
6–7. Sew the sleeve opening.　*8.* Fold the seam allowance toward to cuff.　*9.* Fold the seam allowance of the cuff with an iron.

10

把束带对折。

11

如果搭配蕾丝，可以涂抹布胶。

12

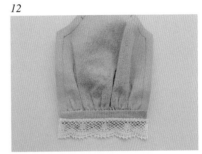

临时固定蕾丝。

13

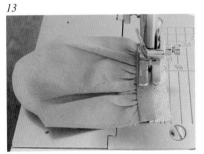

缝合束带。

14

缝合完成的状态。

15

↗ 抽拉

在肩袖的缝头标志处缝出2mm宽的针脚，用于制作褶边。依照大身上肩袖的长度，把褶边聚拢到一起，将线打结。

16

将大身和袖子反面对齐。肩袖的缝头和袖筒缝头仔细对齐，用缝纫机压住，分几次缓慢缝合。

17

将大身反面对齐，沿着袖口、两侧、下摆的顺序缝合。在两侧的缝头处剪出小切口。

18

翻回正面，完成。

10. Fold the seam allowance in half. 11. If you want to put the lace, apply fabric glue. 12. Temporarily the lace. 13–14. Sew the cuffs. 15. Machine sew one line of gathering stitches 2mm in length in the seam allowance. Gather the shoulders until the width fits the armhole. 16. Match the side edge of the sleeve to the bodice and gradually, sew the shoulder of the sleeve to armhole. Raise the machine presser foot a number of times while sewing to gradually align the seam allowances of the sleeve caps and armholes. 17. Place the seam allowance toward the sleeves and iron. With the right sides of the front and back bodice facing. Sew them together. 18. Cut slits into the seam allowance of the pits. Turn right side out. Iron open the seam allowance.

下摆部分调整

l.
Lace Gather
蕾丝褶边

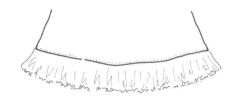

将蕾丝直接缝合在下摆的内侧，不需要制作褶边，只要直接缝合即可。

1

测量想要搭配褶边位置的长度，准备长度约2倍的蕾丝。

2

用单股线缝出一条宽度为 2mm 的针脚，用于稍后制作褶边。
→褶边请参考 p.100

3

依照下摆的宽度聚拢褶边，用熨斗熨烫平整。

4

用熨斗把下摆的缝头折起来。

5

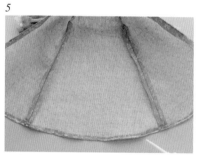

涂抹布胶。

6

聚拢褶边，临时固定蕾丝。

7

缝合。

8

完成。

1. Prepare a lace that is about twice as long as the place where you want to gather the lace gathers.
2. Machine sew one line of gathering stitches 2mm in length in the upper. [refer to p.100 for gathering]
3. Gather the lace until the width fits the hem. Iron flat.　*4.* Fold the seam allowance of the hem with an iron.
5. Apply fabric glue.　*6.* Temporarily the lace on the hem.　*7.* Sew the hem.　*8.* Done.

下摆部分调整

m. *Frill*

褛边

只把布料上部的缝头折起来，聚拢成单侧的褛边。即使从正面缝合也很可爱。

1

用熨斗把布料的两端和下摆的缝头折起来。

2

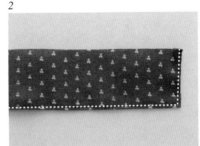

缝合。

3

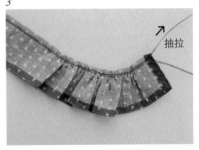

抽拉

用双股线缝出一条宽度为2mm的针脚，用于稍后制作褛边。

→褛边请参考 p.100

4

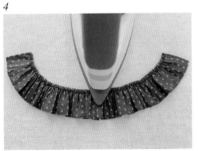

依照下摆的宽度聚拢褛边，将线打结，用熨斗熨烫平整。

5

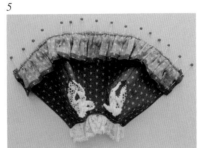

将褛边反面对齐，用别针固定。

6

缝合。

7

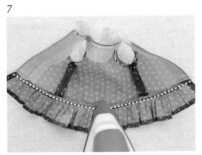

用熨斗把缝头向大身侧熨平。

8

用熨斗压住，缝合。

9

完成。

1. Fold the seam allowance of the ruffled with an iron.　*2*. Sew the edges.
3. Use a machine to sew two line of gathering stitch length 2mm on the upper seam allowance. [refer to p.100 for gathering]
4. Gather the fabric to match of the width of the hem. Iron flat.　*5*. Pin the ruffle and hem.　*6*. Sew the hem.
7. Iron the seam allowance toward to bodice.　*8*. Sew reinforced stitches.　*9*. Done.

下摆部分调整

n.
Drawn Work
抽纱刺绣

抽出布料的横丝，用于制作图案的抽纱刺绣工艺。请使用锦纶等纤薄的平织布料。

1

裁剪用于抽纱刺绣的布料时，要比型纸大一圈。

2

从布边抽出几根纵丝。

3

抽出要制作抽纱刺绣图案部位的横丝。

4
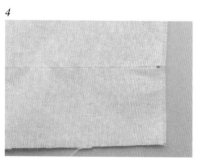
一根一根抽出来。

5
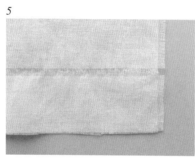
抽出横丝部位约 5mm 宽。

6
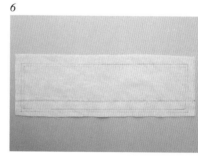
描绘型纸轮廓。

7

剪裁，处理布边的毛茬。

8
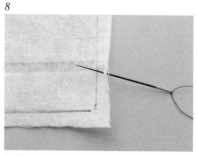
挑起线，从距离布边 2mm 左右的位置入针。

9
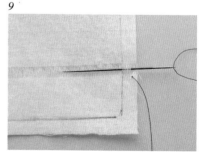
出针后，在正上方挑起约 5mm 宽的纵线，再出针。

1. Cut the dough you want to do drawn work along with the texture, slightly larger than the pattern. *2.* Pull out several warps from the end.
3. Pull the weft thread where you want to apply the drawn work. *4.* Pull the threads one by one.
5. Pull out the weft thread to about 5mm width. *6.* Trace the pattern. *7.* Cut and apply fray stopper liquid to all the edges.
8. Take one thread and pierce the needle about 2mm from the end. *9.* Place your needle behind of the warp threads about 5mm width.

10

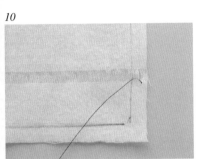

穿过线以后的状态。

11

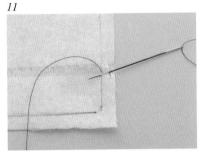

与第9步相同，再次穿针。在约5mm的地方再次出针。

12

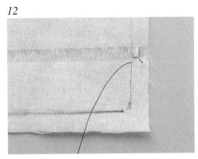

线绕一圈，收紧线。

13

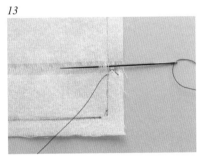

从出针处上面挑起约5mm的纵线。

14

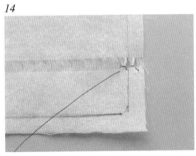

重复步骤9~12。

15

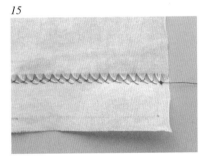

至另一端以后，打球结。从后面看如图所示。

16

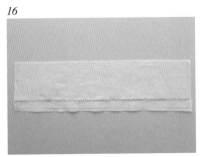

从正面看如图所示。

17

在围裙上的连衣部分做了抽纱刺绣处理。

18

衬衫袖口的距离短，推荐先从这里开始尝试。

10. Pull the needle through to the front.   *11–12.* Put the needle through to the hole as step *9*. Pull out to the hem.
13–15. Repeat steps *9–12* to the full length.   *16.* This is right side.   *17.* The arrangement for apron dress hem.
18. For the sleeve, they are short and easy to challenge.

Gather

褶边

制作裙子、袖子、泡泡袖时用到的褶边制作方法。

1

调整缝纫机的针脚大小，设定在 2~2.5mm。

2

在开始缝和收尾的时候，要重复回针。针脚落在缝头的正中间。

3

为使两边的线更容易拉动，要多留出 15cm 左右。

4

在第一根线缝合位置的旁边，平行缝合第二根线。

5

上线和下线分开。

6

挑出 2 根上线，抽拉聚拢褶边。如果聚拢距离长，从两侧抽拉。如果距离短，单侧抽拉、单侧打结。

7

褶边达到必要长度以后，上线系紧。下线也同样系紧。相反侧也一样需要打结，固定褶边宽度。

8

调整褶边间隔，用熨斗熨平。

9

褶边完成。如果在意缝头褶边的残留线头，可以把线抽掉。

1. Set the machine to 2.5–3.0mm stitch.   *2.* Do not use back stitch as usual the start or end. Sew once along the edge.
3. Leave about 15cm of thread allowance on each side.   *4.* Sew a second line next to the first in the same way.
5. Separate both upper threads from the lower on each side.   *6.* Pull the upper threads while gathering the fabric.
7. When you have the desired width, knot all threads together on either side.   *8.* Iron the gathering to make it neat.   *9.* Cut away the thread allowance.

按扣

服装完工之前固定按扣。本书作品均使用5mm按扣。

1

服装完工之前固定按扣。本书作品均使用5mm按扣。

2

在一个孔处缝2针。

3

将别针固定在母扣中间凹陷处，如图。

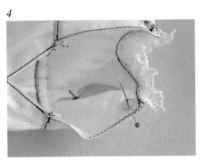

4

将子扣的中心点对应好别针所在位置，如图。

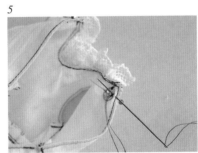

5

带着别针，缝2个孔。

6

取出别针，缝好剩下的孔。

7

缝好下面母扣以后，扣上子扣。确认重合情况，像步骤3那样固定别针。

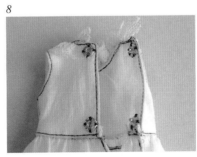

8

如图所示，完成。

1. Start with socket side. Hand sew with two threads. *2.* Insert two needles into each hole.
3–4. Insert a pin all the way through to mark opposite snap placement.
5. Sew through 2 holes with the pin inserted. *6.* Then remove the pin and sew through holes. *7–8.* Attach one more snaps.

─────── 基本款 A 字连衣裙 ───────

M 短款 领子 f, 袖子 j7 分,
双重下摆 m

M 短款 领子 f, 袖子 j7 分,
双重下摆 m

M 中款 / 大身 b, 领子 e, 袖子 j7 分,
下摆 m

M 中款 / 大身 a, 领子 g, 袖子 k, 下摆 l

M 长款 / 领子 g, 袖子 j5 分

S 长款 / 领子 e, 袖子 j3 分

S 短款 / 大身 a, 领子 g, 下摆 m,
袖子 k 袖口搭配蕾丝

L 中款 / 大身 b, 领子 g, 袖子 j7 分

L 长款 / 大身 a, 领子 f, 袖子 k, 下摆 m

─────── 基本款衬衫 ───────

M/ 领子 g, 袖子 h, 下摆 m

M/ 领子 e, 袖子 k,
下摆无褶

S/ 领子 f, 袖子 k

S/ 大身 d, 领子 e, 袖子 i, 下摆 m

L/ 领子 f, 袖子 k

L/ 大身 a, 领子 g, 袖子 i, 下摆 m
袖口处 n

─────── 基本款收腰连衣裙 ───────

M 短款

M 短款 / 领子 f, 袖子 i,
下摆搭配蕾丝

M 中款 / 大身 c, 领子 g, 袖子 j7 分

L 短款 / 大身 d, 领子 e, 袖子
i, 下摆搭配蕾丝

L 长款 / 领子 g, 袖子 j5 分

S 中款 / 大身 a, 领子 g, 袖子
h, 下摆搭配蕾丝

S 短款 / 大身 b, 领子 f, 袖子 j5 分

S 长款 / 领子 g, 袖子 k

S 长款 / 大身 a, 领子 f, 袖子 i,
下摆 n

请从"基本"系列的领子、袖子、下摆等各尺寸型纸中选择制作。

大身部分调整【a 蕾丝拼接 b 蕾丝褶边 c 蕾丝褶皱 d 排褶】

领子部分调整【e 荷叶领 f 圆领 g 蕾丝立领】

袖子部分调整【h 蕾丝鸡翼袖 i 圆袖 j 泡泡袖 k 束带袖】

下摆部分调整【l 蕾丝褶边 m 褶边 n 抽纱刺绣】

荷叶裙

M　　S　　S　　L

连衣围裙

S 长款 / 下摆搭配蕾丝　　M 长款　　M 短款 / 大身 d，下摆 n　　L 短款　　L 长款

灯笼裤

M　　M　　S　　L　　裤腿没有褶边

夹克衫

M/ 泡泡袖　　S/ 长袖，领口没有褶边　　L/ 泡泡袖，领口没有褶边

外套

M/ 长袖，领子有褶边　　S/ 泡泡袖，领子有褶边　　L/ 长袖　　L/ 长袖，特别定制领口

Original Japanese title: DOLL SEWING BOOK「HANON-arrangement-」
Copyright © 2020 SATOMI FUJII
Original Japanese edition published by Hobby Japan Co., Ltd.
Simplified Chinese translation rights arranged with Hobby Japan Co., Ltd. through The English Agency (Japan) Ltd.
and Shanghai To-Asia Culture Co., Ltd.

图书在版编目（CIP）数据

HANON 娃衣缝纫书 /（日）藤井里美著；王春梅
译 . — 沈阳 : 辽宁科学技术出版社 , 2022.1
ISBN 978-7-5591-2157-8

Ⅰ . ① H… Ⅱ . ①藤… ②王… Ⅲ . ①手工艺品－制作
Ⅳ . ① TS973.5

中国版本图书馆 CIP 数据核字（2021）第 153639 号

出版发行 : 辽宁科学技术出版社
　　　　　（地址 : 沈阳市和平区十一纬路 25 号　邮编 : 110003）
印　刷　者 : 辽宁新华印务有限公司
经　销　者 : 各地新华书店
幅面尺寸 : 190mm × 257mm
印　　张 : 6.5
插　　页 : 5
字　　数 : 150 千字
出版时间 : 2022 年 1 月第 1 版
印刷时间 : 2022 年 1 月第 1 次印刷
责任编辑 : 康　倩
封面设计 : 袁　舒
版式设计 : 袁　舒
责任校对 : 徐　跃

书　　号 : ISBN 978-7-5591-2157-8
定　　价 : 79.80 元

联系电话 : 024-23284367
邮购热线 : 024-23284502
E-mail:987642119@qq.com

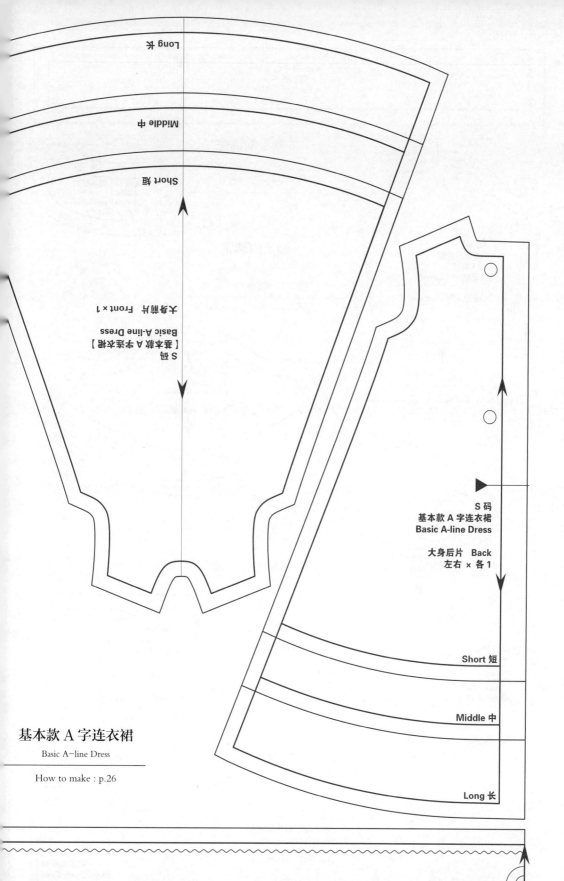

Long 长

Middle 中

Short 短

大身前片 Front × 1

S 码
【基本款 A 字连衣裙】
Basic A-line Dress

S 码
基本款 A 字连衣裙
Basic A-line Dress

大身后片 Back
左右 × 各 1

Short 短

Middle 中

基本款 A 字连衣裙
Basic A-line Dress

How to make : p.26

Long 长

款 A 字连衣裙】
A-line Dress

下摆的调整【m 褶边】
m.Frill × 1

对折 Fold

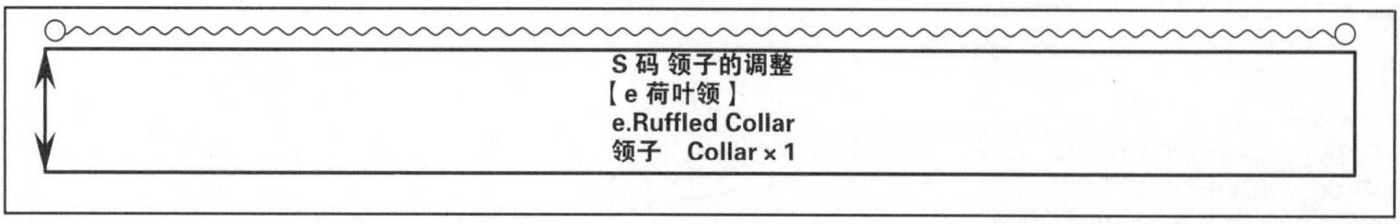

S 码 领子的调整
【e 荷叶领】
e.Ruffled Collar
领子　Collar × 1

领子的调整

Collar Arrangement

e 荷叶领 / *f* 圆领

How to make : p.84,86

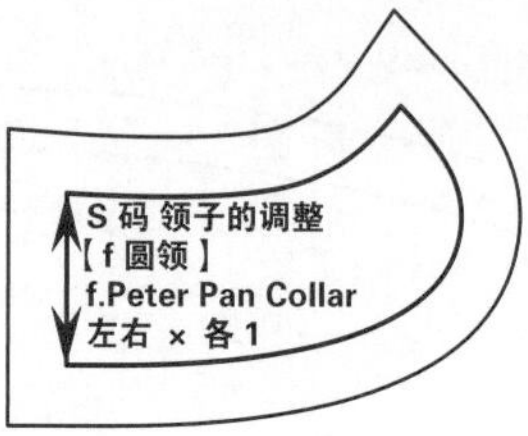

S 码 领子的调整
【f 圆领】
f.Peter Pan Collar
左右 × 各 1

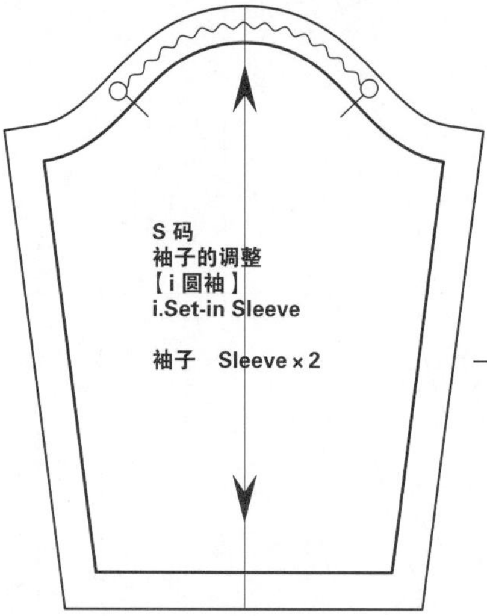

袖子的调整

Sleeve Arrangement

i 长袖 / *j* 泡泡袖 / *k* 束带袖

How to make : p.90,92,94

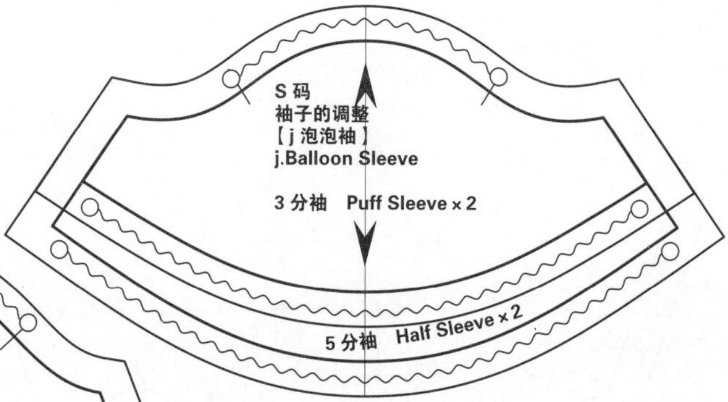

S 码
袖子的调整
【i 圆袖】
i.Set-in Sleeve

袖子　Sleeve × 2

S 码
袖子的调整
【j 泡泡袖】
j.Balloon Sleeve

3 分袖　Puff Sleeve × 2

5 分袖　Half Sleeve × 2

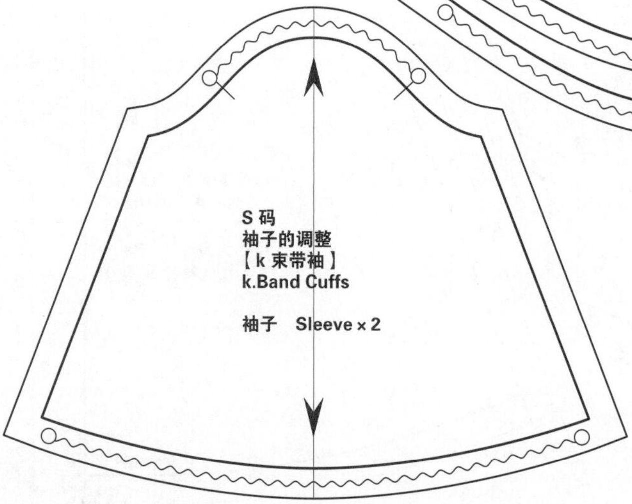

S 码
袖子的调整
【k 束带袖】
k.Band Cuffs

袖子　Sleeve × 2

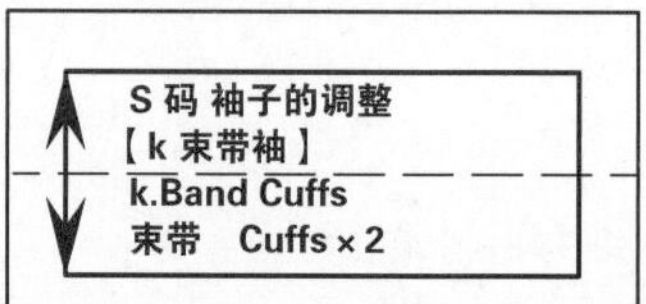

S 码 袖子的调整
【k 束带袖】
k.Band Cuffs
束带　Cuffs × 2

基本款收腰连衣裙

Basic Darts Dress

How to make : p.34

S 码
【基本款收腰连衣裙】
Basic Darts Dress

大身前片　Front × 1

S 码
【基本款
收腰连衣裙】
Basic Darts Dress

大身后片　Back
左右 × 各 1

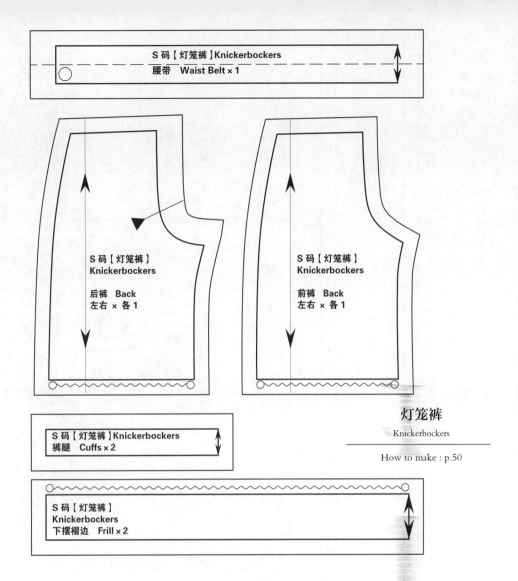

S 码【灯笼裤】Knickerbockers
腰带　Waist Belt × 1

S 码【灯笼裤】
Knickerbockers

后裤　Back
左右 × 各 1

S 码【灯笼裤】
Knickerbockers

前裤　Back
左右 × 各 1

S 码【灯笼裤】Knickerbockers
裤腿　Cuffs × 2

S 码【灯笼裤】
Knickerbockers
下摆褶边　Frill × 2

灯笼裤
Knickerbockers

How to make : p.50

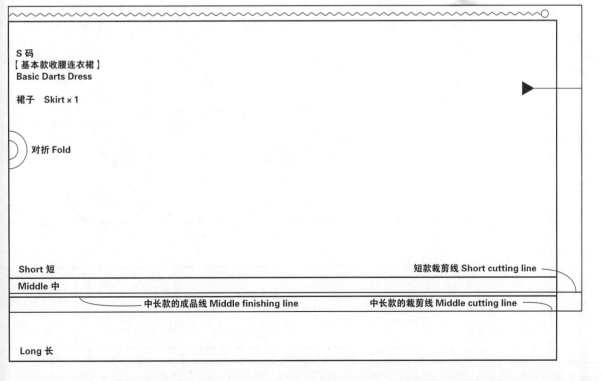

S 码
【基本款收腰连衣裙】
Basic Darts Dress

裙子　Skirt × 1

对折 Fold

Short 短

短款裁剪线 Short cutting line

Middle 中

中长款的成品线 Middle finishing line　　中长款的裁剪线 Middle cutting line

Long 长

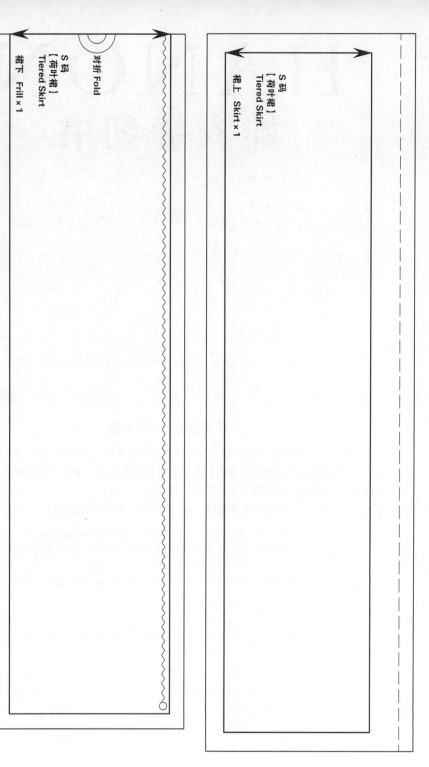

对折 Fold

S 码
【荷叶裙】
Tiered Skirt

裙下　Frill × 1

S 码
【荷叶裙】
Tiered Skirt

裙上　Skirt × 1

荷叶裙
Tiered Skirt

How to make : p.42

HANON

娃衣缝纫书

型纸

型纸

S 码型纸为红色

M 码型纸为黑色

L 码型纸为蓝色

请根据以上颜色分类，选择所需尺码的型纸复印，剪切使用。

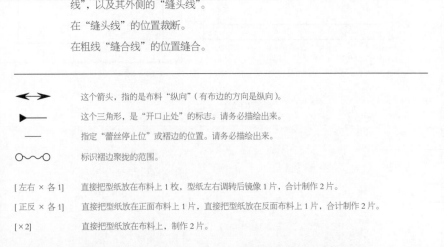

复写型纸的方法

把布料内面放在型纸上面，用粉笔或描图纸等描绘出粗线条的"缝合线"，以及其外侧的"缝头线"。

在"缝头线"的位置裁断。

在粗线"缝合线"的位置缝合。

⟷	这个箭头，指的是布料"纵向"（有布边的方向是纵向）。
▶	这个三角形，是"开口止处"的标志。请务必描绘出来。
—	指定"蕾丝停止位"或褶边的位置。请务必描绘出来。
○〜○	标识褶边聚拢的范围。
[左右 × 各 1]	直接把型纸放在布料上 1 枚，型纸左右调转后镜像 1 片，合计制作 2 片。
[正反 × 各 1]	直接把型纸放在正面布料上 1 片，直接把型纸放在反面布料上 1 片，合计制作 2 片。
[×2]	直接把型纸放在布料上，制作 2 片。

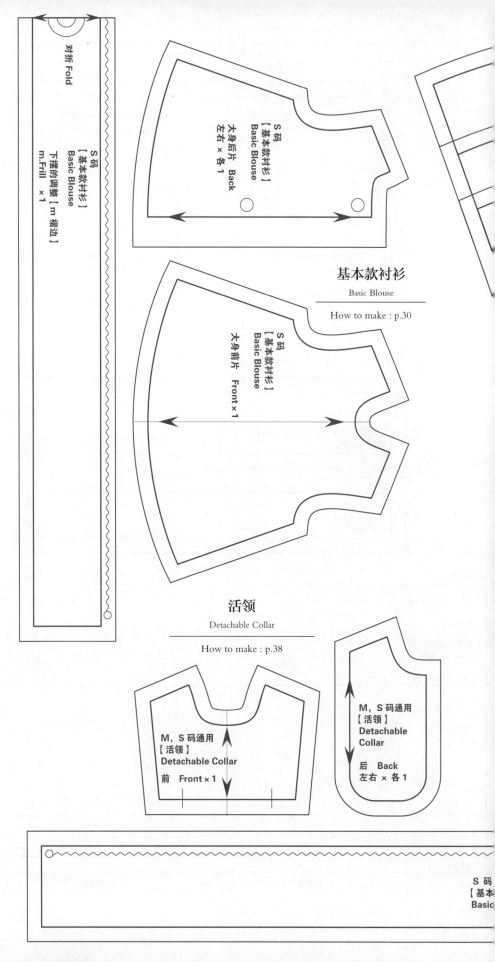

对折 Fold

S 码
【基本款衬衫】
Basic Blouse
下摆的调整［m 褶边］
m.Frill × 1

S 码
【基本款衬衫】
Basic Blouse
大身后片 Back
左右 × 各 1

S 码
【基本款衬衫】
Basic Blouse
大身前片 Front × 1

基本款衬衫

Basic Blouse

How to make : p.30

活领

Detachable Collar

How to make : p.38

M，S 码通用
【活领】
Detachable Collar
前 Front × 1

M，S 码通用
【活领】
Detachable Collar
后 Back
左右 × 各 1

S 码
【基本
Basic

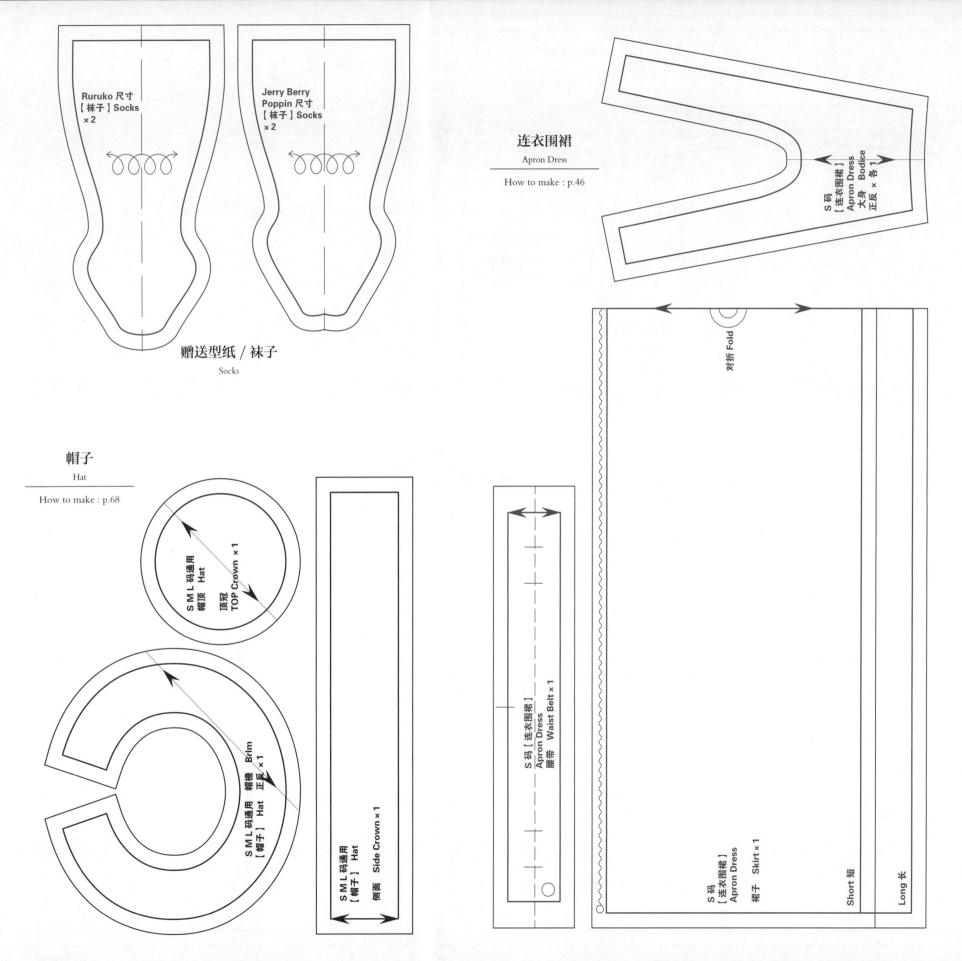

赠送型纸 / 袜子
Socks

Ruruko 尺寸
【袜子】Socks
×2

Jerry Berry
Poppin 尺寸
【袜子】Socks
×2

连衣围裙
Apron Dress

How to make : p.46

S 码【连衣围裙】
Apron Dress　Bodice × 各 1
大身
正反

对折 Fold

S 码【连衣围裙】
Apron Dress　Waist Belt × 1
腰带

S 码【连衣围裙】
Apron Dress　Skirt × 1
裙子

Short 短

Long 长

帽子
Hat

How to make : p.68

S M L 码通用　Hat
帽顶　TOP Crown × 1

S M L 码通用　Hat
帽檐　Brim 正反 × 1

S M L 码通用【帽子】Hat
侧面　Side Crown × 1

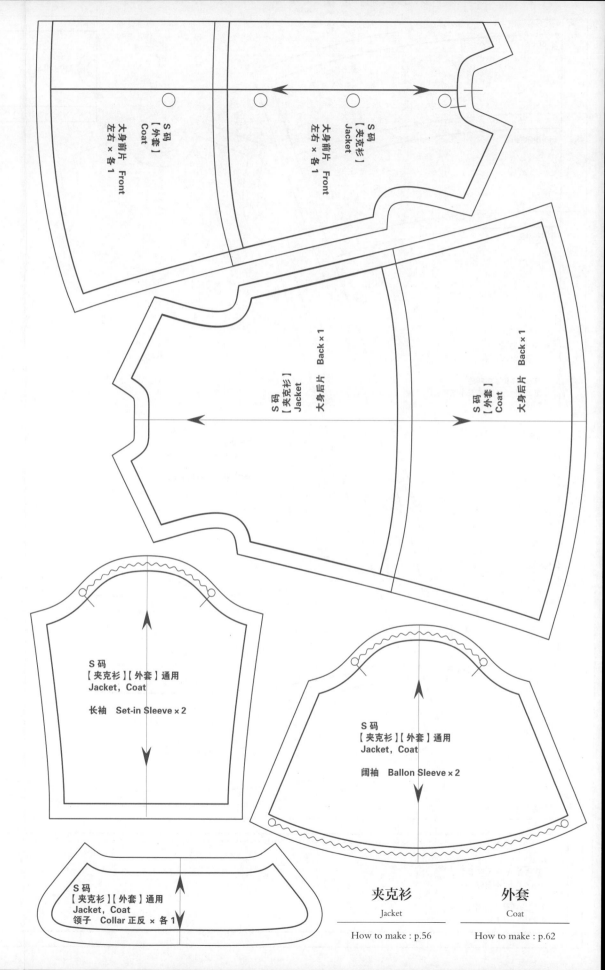

S 码【夹克衫】Jacket
大身前片　Front
左右 × 各 1

S 码【外套】Coat
大身前片　Front
左右 × 各 1

S 码【夹克衫】Jacket
大身后片　Back × 1

S 码【外套】Coat
大身后片　Back × 1

S 码【夹克衫】【外套】通用
Jacket, Coat
长袖　Set-in Sleeve × 2

S 码【夹克衫】【外套】通用
Jacket, Coat
阔袖　Ballon Sleeve × 2

S 码【夹克衫】【外套】通用
Jacket, Coat
领子　Collar 正反 × 各 1

夹克衫
Jacket

外套
Coat

How to make : p.56　　How to make : p.62

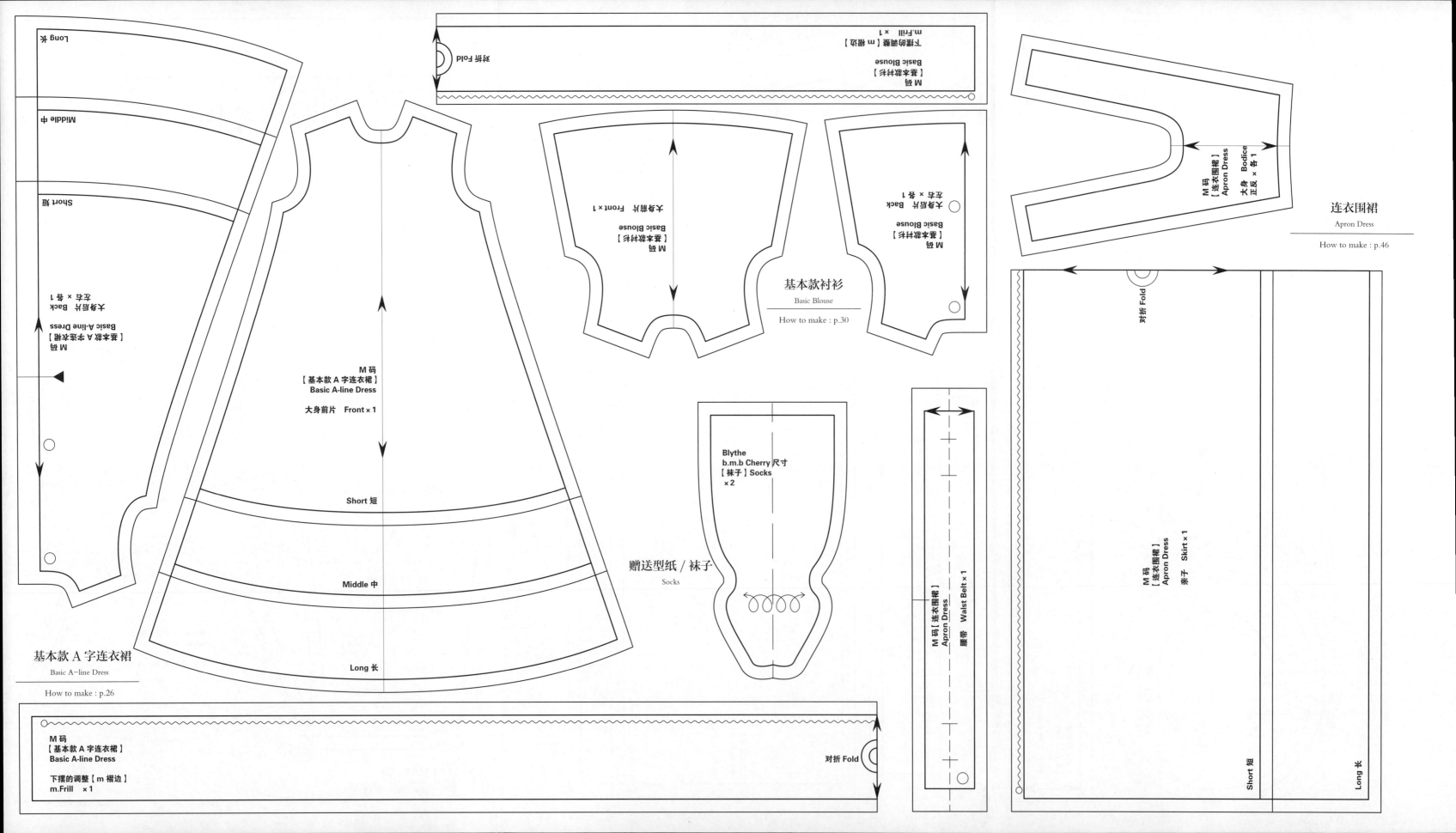

Long 长

Middle 中

Short 短

Basic Blouse
【基本款衬衫】
M 码
m.Frill × 1

对折 Fold

连衣围裙
Apron Dress

How to make：p.46

M 码
【连衣围裙】
Apron Dress
大身 Bodice
正反 × 各 1

大身后片　Back × 1
基本款衬衫
【基本款衬衫】
M 码
Basic Blouse

大身前片　Front × 1
【基本款衬衫】
M 码
Basic Blouse

基本款衬衫
Basic Blouse

How to make：p.30

大身后片　Back × 1
基本款 A 字连衣裙
【基本款 A 字连衣裙】
M 码
Basic A-line Dress

M 码
【基本款 A 字连衣裙】
Basic A-line Dress

大身前片　Front × 1

Short 短

Middle 中

Long 长

Blythe
b.m.b Cherry 尺寸
【袜子】Socks
× 2

赠送型纸 / 袜子
Socks

对折 Fold

M 码
【连衣围裙】
Apron Dress

裙子　Skirt × 1

M 码【连衣围裙】
Apron Dress

腰带　Waist Belt × 1

Short 短

Long 长

基本款 A 字连衣裙
Basic A-line Dress

How to make：p.26

M 码
【基本款 A 字连衣裙】
Basic A-line Dress

下摆的调整【m 褶边】
m.Frill　× 1

对折 Fold

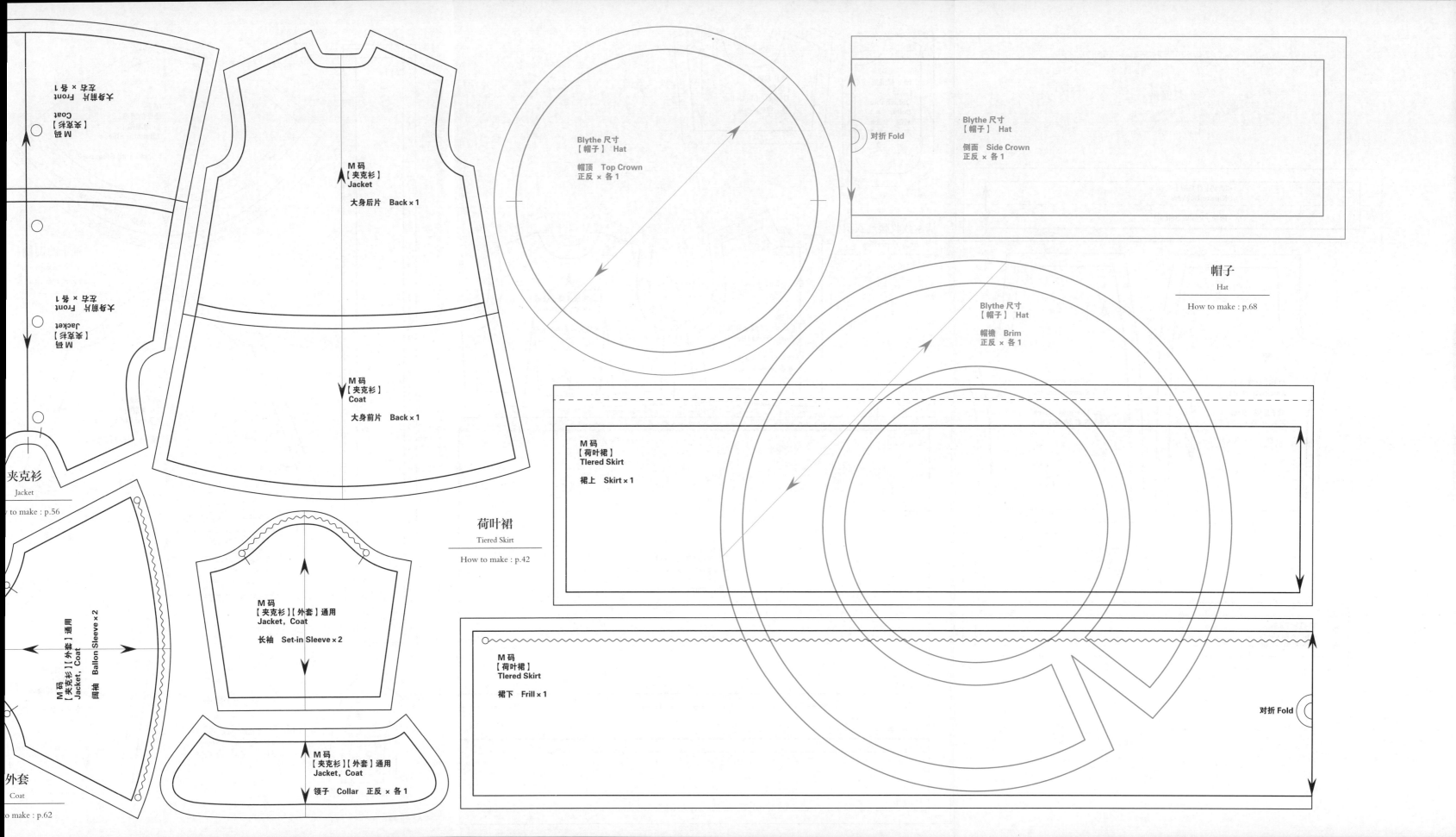

大身后片　Back × 1

M 码
【夹克衫】
Jacket

M 码
【夹克衫】
Coat

大身前片　Back × 1

Blythe 尺寸
【帽子】　Hat

帽顶　Top Crown
正反 × 各 1

Blythe 尺寸
【帽子】　Hat

侧面　Side Crown
正反 × 各 1

对折 Fold

帽子
Hat

How to make : p.68

Blythe 尺寸
【帽子】　Hat

帽檐　Brim
正反 × 各 1

M 码
【荷叶裙】
Tiered Skirt

裙上　Skirt × 1

荷叶裙
Tiered Skirt

How to make : p.42

M 码
【荷叶裙】
Tiered Skirt

裙下　Frill × 1

对折 Fold

夹克衫
Jacket

How to make : p.56

M 码
【夹克衫】【外套】通用
Jacket, Coat

阔袖　Ballon Sleeve × 2

M 码
【夹克衫】【外套】通用
Jacket, Coat

长袖　Set-in Sleeve × 2

M 码
【夹克衫】【外套】通用
Jacket, Coat

领子　Collar　正反 × 各 1

外套
Coat

How to make : p.62

灯笼裤
Knickerbockers

How to make : p.50

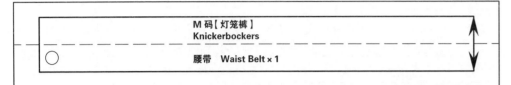

M 码【灯笼裤】
Knickerbockers
腰带　Waist Belt × 1

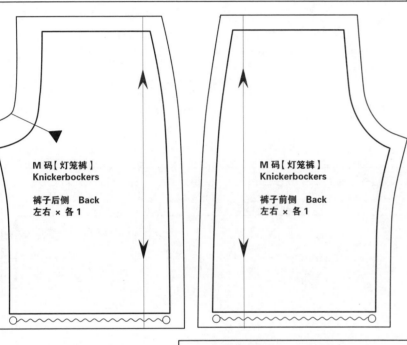

M 码【灯笼裤】
Knickerbockers
裤子后侧　Back
左右 × 各 1

M 码【灯笼裤】
Knickerbockers
裤子前侧　Back
左右 × 各 1

M 码【灯笼裤】
Knickerbockers
裤腿　Cuffs × 2

M 码【灯笼裤】
Knickerbockers
裤腿褶边　Frill × 2

M S 码通用
【活领】
Detachable
Collar
后侧　Back
左右 × 各 1

M S 码通用
【活领】
Detachable Collar
前　Front × 1

活领
Detachable Collar

How to make : p.38

基本款收腰连衣裙
Basic Darts Dress

How to make : p.34

M 码
【基本款收腰连衣裙】
Basic Darts Dress
大身后片　Back
左右 × 各 1

M 码
【基本款收腰连衣裙】
Basic Darts Dress
大身前片　Front × 1

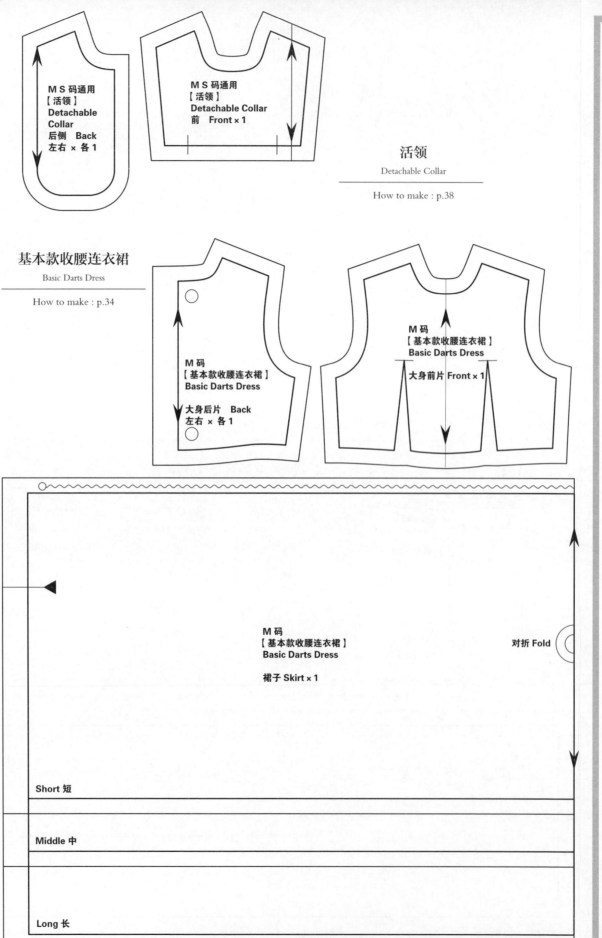

M 码
【基本款收腰连衣裙】
Basic Darts Dress
裙子 Skirt × 1

对折 Fold

Short 短

Middle 中

Long 长

M 码　领子的调整
【e.荷叶领】
e.Ruffled Collar
领子　Collar × 1

M 码【领子的调整】
【f.圆领】
f.Peter Pan Collar
左右 × 各 1

领子的调整
Collar Arrangement
e 荷叶领 /f 圆领

How to make : p.84,86

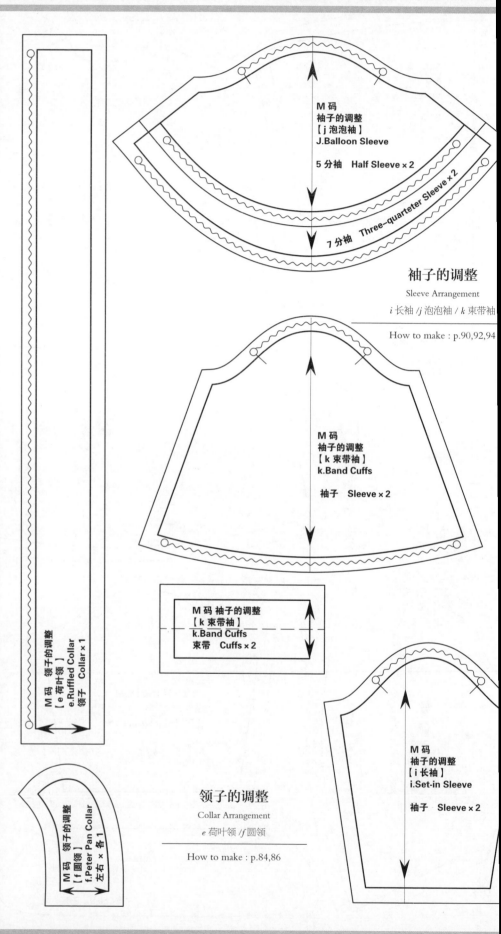

M 码
袖子的调整
【j 泡泡袖】
J.Balloon Sleeve
5 分袖　Half Sleeve × 2
7 分袖　Three-quarteter Sleeve × 2

袖子的调整
Sleeve Arrangement
i 长袖 /j 泡泡袖 / k 束带袖

How to make : p.90,92,94

M 码
袖子的调整
【k 束带袖】
k.Band Cuffs
袖子　Sleeve × 2

M 码 袖子的调整
【k 束带袖】
k.Band Cuffs
束带　Cuffs × 2

M 码
袖子的调整
【i 长袖】
i.Set-in Sleeve
袖子　Sleeve × 2

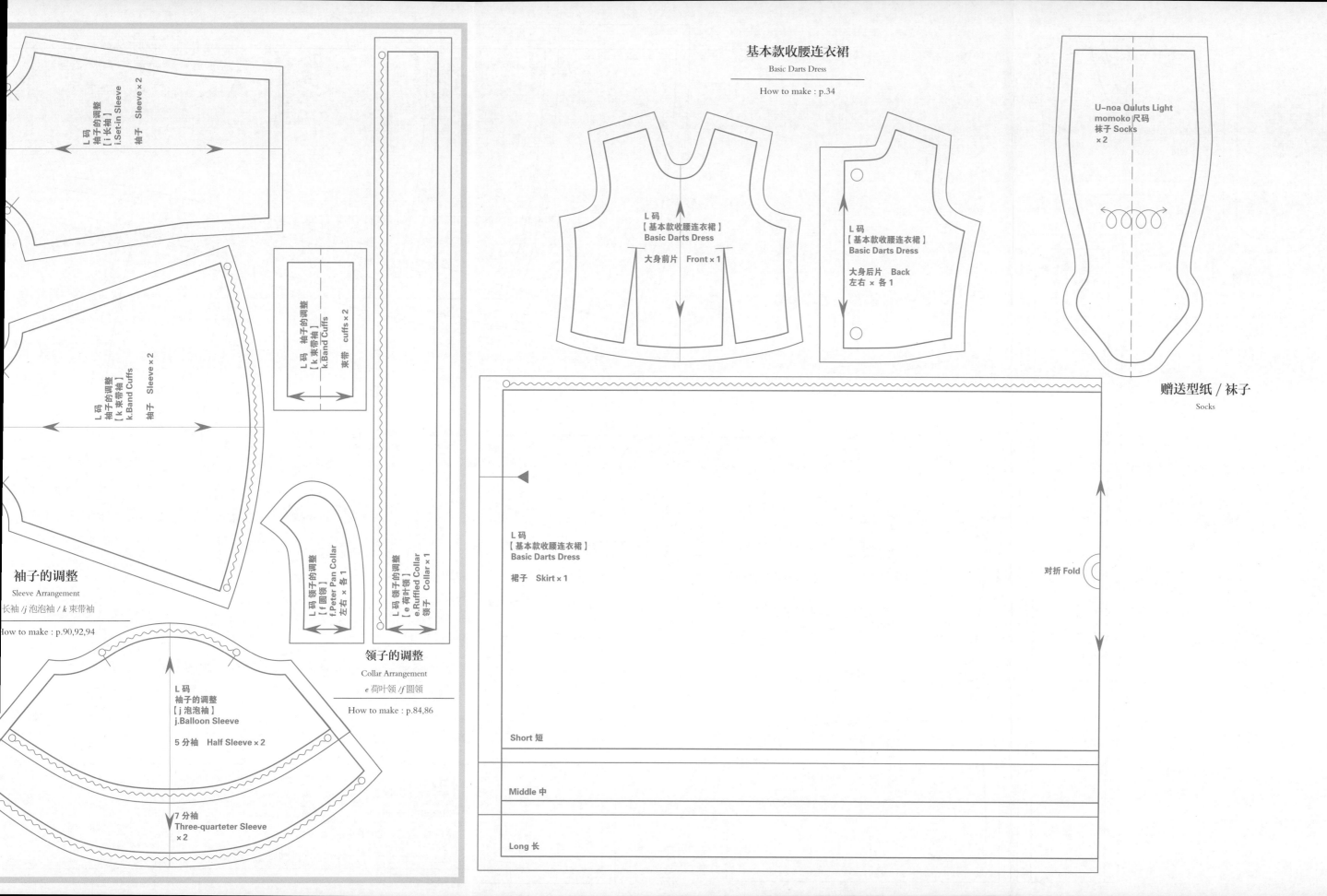

基本款收腰连衣裙
Basic Darts Dress

How to make : p.34

L 码
袖子的调整
[i 长袖]
i.Set-in Sleeve

袖子　Sleeve × 2

L 码
袖子的调整
[k 束带袖]
k.Band Cuffs

袖子　Sleeve × 2

L 码 袖子的调整
[k 束带袖]
k.Band Cuffs

束带　cuffs × 2

袖子的调整

Sleeve Arrangement

长袖 /j 泡泡袖 / k 束带袖

How to make : p.90,92,94

L 码
袖子的调整
[j 泡泡袖]
j.Balloon Sleeve

5 分袖　Half Sleeve × 2

7 分袖
Three-quarteter Sleeve
× 2

L 码 领子的调整
[f 圆领]
f.Peter Pan Collar
左右 × 各 1

L 码 领子的调整
[e 荷叶领]
e.Ruffled Collar
领子　Collar × 1

领子的调整

Collar Arrangement

e 荷叶领 /f 圆领

How to make : p.84,86

L 码
【基本款收腰连衣裙】
Basic Darts Dress

大身前片　Front × 1

L 码
【基本款收腰连衣裙】
Basic Darts Dress

大身后片　Back
左右 × 各 1

U-noa Quluts Light
momoko 尺码
袜子 Socks
× 2

赠送型纸 / 袜子

Socks

L 码
【基本款收腰连衣裙】
Basic Darts Dress

裙子　Skirt × 1

对折 Fold

Short 短

Middle 中

Long 长

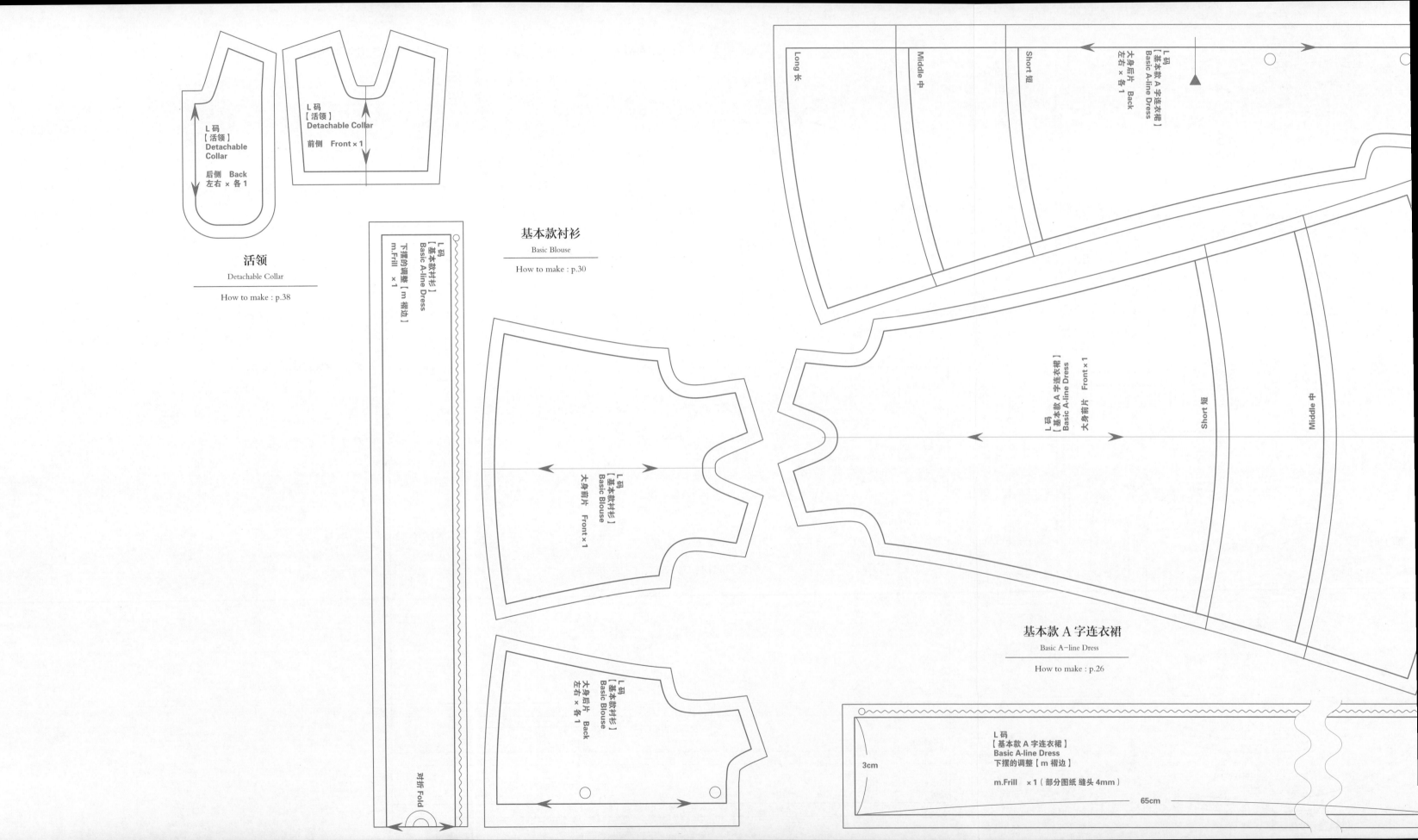

活领
Detachable Collar

How to make : p.38

基本款衬衫
Basic Blouse

How to make : p.30

基本款 A 字连衣裙
Basic A–line Dress

How to make : p.26